The DISCOVERY of the GERM

JOHN WALLER

Series Editor: Jon Turney

REVOLUTIONS IN SCIENCE

Published by Icon Books UK

Originally published in 2002 by Icon Books Ltd.

This edition published in the UK in 2004
by Icon Books Ltd., Grange Road,
Duxford, Cambridge CB2 4QF
E-mail: info@iconbooks.co.uk
www.iconbooks.co.uk

Sold in the UK, Europe, South Africa
and Asia by Faber and Faber Ltd.,
3 Queen Square, London WC1N 3AU
or their agents

Distributed in the UK, Europe, South Africa
and Asia by TBS Ltd., Frating Distribution Centre,
Colchester Road, Frating Green, Colchester CO7 7DW

This edition published in Australia in 2004
by Allen & Unwin Pty. Ltd.,
PO Box 8500, 83 Alexander Street,
Crows Nest, NSW 2065

Distributed in Canada by
Penguin Books Canada,
10 Alcorn Avenue, Suite 300,
Toronto, Ontario M4V 3B2

ISBN 1 84046 502 6

Series editor: Jon Turney

Originating editor: Simon Flynn

Typesetting by Hands Fotoset

Printed and bound in the UK by Cox and Wyman Ltd., Reading

Cover design and typography by Nicholas Halliday

Contents

List of Illustrations

Acknowledgements

I have incurred many debts in writing this book: to the many individual historians upon whose accounts I have drawn, and whose works are listed in the bibliography to this book, to Jon Turney, Simon Flynn, Michael Waller, Richard Graham-Yooll and Lawrence Hill for their excellent editorial advice, and to Alison Stibbe for generously translating key articles. My wife, Abigail, has been a tremendous source of support and it is to her that this book is dedicated.

About the Author

John Waller read Modern History at Oxford University and took Masters degrees in Human Biology and the History of Science and Medicine. He gained his Ph.D. from University College London in 2002 and is now a Research Fellow at UCL's Wellcome Trust Centre for the History of Medicine. He is the author of *Fabulous Science: Fact and Fiction in the History of Scientific Discovery* (2002).

To Abigail

INTRODUCTION: REVOLUTIONARY, BY ANY STANDARDS

Words, like coins, are subject to devaluations and debasement. Big words, like revolution, are particularly vulnerable. Because of its proper coupling with adjectives such as American, French or Industrial, the term 'revolution' has been exploited by a myriad writers seeking to hype up some comparatively minor change in, say, the kitchen, the workplace or the high street. In contrast to such insignificant events, a real revolution is something that transforms major aspects of our world and the way we see it. The extraordinary, albeit bloodless, scientific revolution that took place between 1880 and 1900 provides us with a paradigmatic example. For, in this short space of time, medicine underwent perhaps its greatest ever transformation. In just 20 years, the central role of germs in producing illness was for the first time decisively demonstrated and Western doctors abandoned misconceived ideas about the causes and nature of disease that had persisted, in one form or another, for thousands of years.

This extraordinary revolution was driven forward by two fiercely competitive teams. One was led by the painstaking, systematic and observationally brilliant German scientist, Robert Koch, the other by the bold, risk-taking and fabulously creative Frenchman, Louis Pasteur. The story of how mankind learned that tiny germs are the cause of infectious disease is in large measure an exploration of the remarkable series of experiments performed by this handful of men. Few individuals have ever had such a profound and lasting impact upon any field of human endeavour as these. Nor has science often seen rivalries so bitter and yet so astoundingly fruitful. When Pasteur and Koch embarked on their scientific careers, the germ theory had for centuries been little more than a loose conjecture, dismissed as fanciful by almost the entire medical establishment. In their final years, it was a universally attested fact of medical science.

The medical world of 1900 was, in consequence, utterly different to that of 1800. After millennia of wishful thinking and groping in the dark, medical science at last got it right. Theories that had appeared entirely serviceable a few years before quickly became amusing curiosities. Doctors and surgeons wondered what could have induced their predecessors to bleed their patients till they were barely conscious. Others realised for the first time that coughs and sneezes really do spread diseases. And many more looked back in horror at their younger days, when a quick wipe of the scalpel upon an already bloodied apron, before moving on to the next surgical patient, was deemed ample hygienic precaution. Breathtakingly rapid, the discovery that germs cause disease was emphatically revolutionary.

Anatomy of a Revolution

This book charts how, why and by whom germ theory was transformed from a hotly disputed speculation to a central tenet of modern medicine. The issues at stake are simply stated. For germ theory to become the orthodox view, three things had to be established. First, that microbes can cause illnesses within the body. Second, that they can be spread from one person to another. And, third, that for each form of infectious disease there is a specific microbial agent. In other words, the same microbe will always produce the same disease in susceptible hosts.

Naturally, more than a century after the germ theory was vindicated, these ideas can seem rather obvious. So our first challenge is to get inside the heads of people who for so long resisted a theory that to us is perfectly straightforward. Until around the 1850s, most doctors had always assumed that each disease could be caused in a variety of different ways; people succumbed to exactly the same illnesses but for entirely different reasons. There were no necessary causes and, as a result, there was little impetus for doctors to look for specific disease-causing agents such as germs. This is the medical worldview that germ theory would have to overthrow. Yet, despite this major disparity between medicine old and new, the germ revolution didn't appear out of nowhere during the late 1800s. Much as Sir Isaac Newton spoke of having stood upon 'the shoulders of giants', the great names of the germ revolution owed much to those who preceded them.

To trace the origins of this revolution, we need to go back to the invention of the microscope and to the pioneers who were stunned by the realisation that nearly every square inch of the world teems with microbial life. In the following centuries, the catastrophic spread of giant epidemic killers, the rise of an increasingly scientific medicine and radical changes in the nature of the doctor–patient relationship transformed the medical profession's attitudes to disease and, eventually, allowed them to speculate on the role of germs in causing it. Finally, between 1880 and 1900, an explosive burst of experimental activity at last drove home the truth of germ theory to all but the most purblind of critics. And it is the sheer pace, intensity and excitement of these crucial decades that demands the adjective 'revolutionary'.

Pasteur, Koch and their less famous supporters, hundreds of human guinea pigs and countless unremembered laboratory animals, all played essential roles in the germ revolution. But while this is a story rich in individual genius, self-sacrifice and experimental virtuosity, stripping away more than a century's accretion of romantic myth reveals that its triumphant conclusion owed as much to luck and raw ambition as it did to investigative brilliance and humanitarian resolve. Yet even if recent scholars have worked hard to demythologise the history of germ theory, not even modern historians deny that this was a revolution in the very fullest sense of the term.

Part I

Before the Germ

· CHAPTER 1 ·

THE WORLD ACCORDING TO WILLIAM BROWNRIGG

We start with an extract taken from the medical casebook of the eighteenth-century English physician William Brownrigg. A cultivated and learned man, trained in the best medical academies of the time, Brownrigg was at the leading edge of medical science. On 13 November 1738, he was called to attend to a 'spotty, delicate girl' called Miss Musgrave, who was suffering from a serious fever. 'Her face', Brownrigg later noted,

> *was puffed up into a swelling, which first appeared on her forehead and then spread downward to her nose, upper lips and cheeks. It was attended with great pain and her urine was pale ... She was bled seven times within six days and a large quantity was obtained each time, so that the patient often felt faint.*

Then, Brownrigg continued, with the use of 'local plasters applied to the back of the neck and lower legs, nitrous powders and tartar, and a suitable cooling diet, the disease completely cleared up'.

We can now be fairly confident that Miss Musgrave had a nasty bacterial infection called erysipelas, from which she was lucky to recover. But while Brownrigg was happy to attribute her survival to his care, he obviously had no conception that micro-organisms were the cause of the condition. Instead, he blamed Miss Musgrave's fever on her 'delicate constitution', a build-up in her body of 'peccant humours' and the fact that the weather had been 'excessively wet & rainy and moist and cold with Westerly Winds'. William Brownrigg was too well educated and too upstanding a member of his local community to have been a charlatan. So how can one make sense of this frankly bizarre diagnosis and treatment?

Medicine's Sense of Humour

In the venerable tradition of Hippocrates and Galen, eighteenth-century doctors saw illness as a deviation from a state of health, caused by the violation of natural laws. These laws took into account a wide range of environmental, physical and psychological factors known as the 'non-naturals', including air, food and drink, movement and repose, sleeping and waking, excretion and retention, as well as a person's state of mind. Whenever a disharmony arose between any of these non-naturals and the individual's physical being, ill health was the inevitable result. For example, poor quality air, an excess of venous spirits, a melancholy frame of mind, suppressed sweating, even an overly sedentary lifestyle could all be seen as the direct causes of what we now know to be infectious disease.

Again drawing on the ancients, physicians of the 1700s believed that the non-naturals caused illness by disturbing the body's fluids or humours, whether blood, phlegm, bile, urine, sweat or something else. A person became unwell when an oversupply of one of their fluids produced a disequilibrium or when they became corrupted or 'peccant'. So if a patient developed a build-up of phlegm, most physicians unhesitatingly identified this as the underlying disease and its removal the only possible cure. By helping to evacuate the phlegm, the physician saw himself as aiding the body in restoring a proper balance of its humours or ridding it of noxious fluids.

Take, for instance, Brownrigg's description of the case of a nobleman suffering from erysipelas. The patient's fever indicated to Brownrigg that his body's attempts to expel noxious or excess fluids were being thwarted by an internal blockage. To overcome this, the nobleman was prescribed a heady cocktail of 'mercury dissolved in wine', the effect of which was to cause his sinuses to start expelling large quantities of phlegm. A few days later, Brownrigg felt able to proclaim the verdict 'cured completely'. The medicine, he explained, had strengthened the 'expulsive faculty hence the material of erysipelas was removed by blowing of the nose'.

This kind of logic explains the medical profession's widely remarked-upon obsession with the texture of blood, and the odour, consistency and colour of the patients' stools. It also accounts for the rich armoury of emetics, cathartics, diuretics and diaphoretics to which their patients were subjected. Suffice to say, 'peccant' and 'balance' were among the three most popular terms in the medical lexicon. The other was 'inflammation'. Many physicians saw the body as a kind of hydraulic machine, with its veins, arteries and pores analogous to the pipes, valves, pumps and ducts used in water mills. The non-naturals, they argued, could also disturb these solid components of the body, causing swelling and impeding the free flow of the humours. These then built up and became poisonous, producing anything from typhoid to scurvy. Reducing inflammation was thus another major preoccupation of Brownrigg's age.

Causes and Effects

A fundamental feature of the humoural theory, and the variants of it that survived into the nineteenth century, is that there was no such thing as a specific disease. The precise form an illness took was seen to be dependent on the humours involved, the place in the body where they had accumulated, and the site at which the body was seeking to expel them. Since all

Illustration 1: A typical eighteenth-century medical encounter: a woman being bled by a surgeon as she is comforted by a female friend. Coloured etching by Thomas Rowlandson (1756–1827). Source: The Wellcome Library, London.

these were highly unpredictable, few physicians spoke of diseases having predictable courses. Naturally, they could tell the difference between measles, plague and scarlet fever. But there was always an expectation that one condition, through bad luck or improper treatment, would suddenly turn into another.

Physicians and patients alike fretted that 'peccant humours' might quit one part of the body and settle in a much more delicate area. 'By drinking too freely of cooling Liquers in order to dilute my Blod and put off the Gout,' wrote one William Abel in 1718,

'I flung myselfe into a diabetes, much the more dangerous distemper of the two.' A century and a half later, we find the same idea in the writings of the English heroine, Florence Nightingale. 'I have seen diseases begin, grow up and pass into one another. The specific disease doctrine', she harangued, 'is the grand refuge of weak, uncultured and unstable minds.'

Another striking feature of this framework of ideas is that doctors very seldom relied on mono-causal explanations of illness. In explaining outbreaks of, say, food poisoning or influenza, today we wouldn't look much further than the viruses or bacteria responsible. In contrast, eighteenth-century notions of causality were nearly always pluralistic. A good example is Brownrigg's account of the causes of haemorrhoids. As a sufferer himself, Brownrigg did lots of ruminating on this subject, but there was nothing unconventional about his description:

> *bad digestion, arising from a strong strain of melancholy humour, which often affects those who use thick foods, hard to digest, who wear themselves out with strenuous drinking bouts or who are weighed down all day with cares and sadness or live a sedentary life or, finally, those who apply themselves too earnestly to their studies, especially at night.*

Physicians like Brownrigg also divided up the various causes of disease into 'predisposing' and 'exciting' factors, both of which were needed to cause ill health. 'Predisposing' causes usually had to do with the pre-existing state of the individual's humoural constitution, the prevailing climate and the quality of the air they breathed. 'Exciting' causes encompassed such things as poisonous fumes floating in the atmosphere (usually known as miasmas), periods of mental anxiety and just about any form of over-indulgence.

Crucially, these notions of exciting and predisposing cause dispensed with the need to think in terms of specific diseases with specific causes. Rather, entirely different ailments were often seen as each person's individual response to the same noxious agent. Nearly all doctors assumed, for instance, that those who inhaled noxious fumes succumbed to diseases like cholera, typhoid, diphtheria and dysentery. But the particular sickness developed was felt to depend on the person's own history and susceptibilities, and not on the type of poison ingested.

Conversely, where two patients had identical symptoms, doctors often invoked a very different combination of causal factors. For instance, according to William Buchan's 1774 best-seller *Domestic Medicine*, a factory labourer with scurvy would be told that his rotten gums, painful joints, tiredness and ulcerations were the result of 'vitiated humours' caused by poor clothing, a lack of personal hygiene and his unwholesome diet. In contrast, a lord of the manor with scurvy would probably be chastised for eating too much rich and hard-to-digest food and spending far too much time sitting in his armchair rather than being outside inhaling pure air. The cause of illness was in each case a matter of individual lifestyle.

To summarise, eighteenth-century ideas about the cause and the cure of ill health all rested on a fundamental belief that disease results from a disharmony between the individual's physiological state and their mode of life. And if illness was seen to be an individual response to unhealthy lifestyles, it makes sense that physicians tailored their diagnosis and treatment to each client. Ways of combating their unique predispositions were therefore combined with appropriate dietary tips and practical methods of removing their noxious or over-abundant bodily fluids.

So when Brownrigg assessed what was wrong with the sickly Miss Musgrave, he wasn't hedging his bets by producing a long list of causes. On the contrary, thoroughly in keeping with contemporary medical thought, he first presented a credible predisposing cause, her 'sickly constitution', then an exciting cause, 'excessively wet & rainy' weather and 'Westerly Winds' and, finally, he deduced the diseased state caused by these conditions, 'a build up of peccant humours'. Next, and again according to the wisdom of the time, Brownrigg made every effort to remove the toxic fluids from her body and gave her a cooling diet to reduce inflammation. In short, Miss Musgrave's parents had every reason to commend themselves for selecting a skilful and learned physician.

Good Science, Bad Medicine

If, as the above account suggests, Brownrigg and his colleagues were neither rogues nor fools, why didn't they so much as speculate on the role of germs in causing disease? One answer to this question is that the evidence for germ theory remained very weak until the late nineteenth century. But it is also important to appreciate the many strengths of Brownrigg's medical worldview. Like all good scientific theories, his provided a simple explanation for a vast number of distinct physical phenomena. Coherent and easy to visualise, the idea of vitiated and imbalanced humours made complete sense of most aspects of health and disease.

After all, many sicknesses *are* strongly correlated with the production, expulsion and retention of rather unpleasant bodily fluids. If you vomit as a consequence of food poisoning, the body expels bile with the offending material. If you have tuberculosis, you cough up bloody sputum. If you have plague, large lymph-filled buboes develop under the arms and in the region of the groin. Likewise, in most cases, if the vomiting or coughing stops, or the swellings go down and the bumps disappear, the patient has recovered. In the absence of modern knowledge about disease-causing microbes, this kind of observational evidence lent real credence to the idea that disease is nothing but the excessive build-up or corruption of bodily fluids.

The humoural theory also drew strength from its extreme versatility. If therapy failed, despite copious bleeding and almost every conceivable form of induced excretion, then the individual's equilibrium was declared beyond restoration. Conversely, if therapy succeeded, the doctor had yet another confirmation of the veracity of the theories he had been taught. But again, lest this sound like quackery, it has to be recognised that central to the success of humoural theory was its acceptance by doctor and patient alike. Not only did these ideas seem to fit reality, they were also mostly drawn from ancient texts that both doctors and laymen revered. Humouralism survived largely because it was part of the common intellectual heritage of the civilised world.

Bedside Manners

Another reason why Brownrigg's medical worldview persisted is that it mapped so neatly onto the cleavages of eighteenth-century society. In this age of rigid hierarchy, wealthy patients usually called the shots and physicians had to observe strict rules of deference. This subservience took several forms. Most obviously, it was the physician's duty to visit his patients, not vice versa. But, as a result of spending most of their time at the bedside of individual patients, most doctors acquired an oddly individualised impression of the nature of disease. Focusing on single patients and their lifestyles, the peculiarities of their ailments stood out and the common features of illness receded from view. Alas, this narrow focus left physicians much less open to the recognition that many of their patients were suffering from the same illnesses with identical causes.

The same sense of social inferiority discouraged the physician from developing any theory likely to conflict with the common sense and general knowledge of his genteel clients. Where novelty might well be seen as putting on intellectual airs, sticking close to ancient theory, with just a few modern trappings, was much the safer course. Social propriety equally deterred the physician from conducting proper physical examinations. Edinburgh's John Rutherford, for instance, wrote in 1768 of a female patient who explained that her ill health was caused by 'the mismanagement she underwent in childbed'. She says, Rutherford wrote, that 'she was lacerated and probably it was her vagina'. Yet there was no suggestion that he lift her petticoats and take a look for himself. Instead, he used the case to stress the merits of always inspecting the patient's gums and the skin beneath the eyelids.

Where examination was limited to the face, the pulse, bodily fluids and general demeanour, it is not surprising that knowledge of disease was mostly limited to what we now know to be just its symptoms. Yet most physicians were perfectly content with this state of affairs. For them, medicine was a cerebral activity that didn't require much physical contact.

In any case, too much touching threatened to erode the social gap between the educated physician and the lowly, and necessarily more tactile, surgeon (who spent as much time cutting hair as performing operations). Physicians saw little point in exploring more deeply. Nor were they given much encouragement to do so. And even if an opportunity for dissection did come their way, it usually involved hanged felons, who, except for a broken neck and perhaps a few intestinal worms, would usually be in pretty good health.

Still, if rich clients had most of the power in the doctor–patient relationship, they paid a heavy price for it in more than one sense. In order to win loyalty, physicians had to convince patients that their remedies weren't the ineffectual cure-alls available from the multitude of eighteenth-century wise women and quacks. As a result, they came to favour the kinds of heroic therapy that left patients sealed to the privy or bent over a basin salivating uncontrollably. The effects of such 'treatments' were so extreme that fee-paying patients felt that at least they were getting something for their money: the acute discomfort involved, far from alienating patients, served to convince many of them that their man definitely knew what he was about.

This tight fit between social relations and ideas about health and disease meant that it would take a huge shake-up in the structure of society before medical progress became at all likely. It is thus no coincidence that the germ revolution took place against a backdrop of industrial revolution during which eighteenth-century social strata buckled, folded and, in some countries, collapsed. As such, the relentless clatter of the textile mill, the regular chug of the steam train and the strident tones of the radical orator provide a fitting score to this drama.

The More Things Change ...

From the early eighteenth century, however, some parts of the ancient theoretical framework had already come under attack. This assault was partly fuelled by new discoveries. As more was learned about the nervous system, many doctors began to ascribe disease as much to disordered nerves as to humoural flow, imbalance and inflammation. This stress on nerves markedly increased once it had been shown that electric currents could be used to control the motor nerves of frogs and, later, even the facial expressions of guillotined aristocrats whose heads were pulled out of blood-sodden baskets and electrified to amuse the *sans culottes*.

Nevertheless, before the 1820s, such developments never seriously challenged the popularity of humoural theory. Rather than overturning ancient ideas, many doctors simply incorporated new findings into pre-existing frameworks; old templates were continually recycled. So firm was the grip of ancient ways of thinking that, as late as the 1850s, doctors still placed great emphasis on the non-naturals, and bleeding, purging and enemas remained their choice remedies. Above all, early Victorian doctors clung fast to the notion that any single disease could be produced in all manner of different ways.

Yet although early proponents of germ theory were far from pushing at an open door, some factors did strongly favour them. For all the strengths of the humoural worldview, the first seeds of its ultimate destruction had been sown not long after it came into being. More were planted during the Renaissance, and although germination was painfully slow at first, by the early 1800s the crop was beginning to ripen. The next few chapters look at three of the main trends that eventually combined to complete this process and, in so doing, made the germ revolution possible. First, the discovery of the microbe and the gradual recognition of its role in causing fermentation and decay. Second, the transformations in the practice of medicine that arose directly out of the French Revolution. And, lastly, the gradual appreciation of the importance of both infection and contagion.

Part II

The Germs of Revolution, 500 BC–1850 AD

· CHAPTER 2 ·

ARSENALS OF DEATH

The idea, central to modern germ theory, that disease is caused by something 'out there' was no less fundamental to ancient medical tradition. This much is clear from the title of Hippocrates' best-known work, *Airs, Waters and Places*, written in the fifth century BC. To the father of modern medicine, a wide range of environmental factors, from the weather and seasons to altitude and wind direction, were quite capable of causing illness. But Hippocrates singled out towns located near stagnant stretches of water or marshy ground as the places most prone to disease. From these, it was believed, the wind carried deadly fumes to those living in the vicinity. Such noxious gases were loosely defined. But fetid water and filthy conditions are obviously correlated with illness and, relying on the simple diagnostic tool of smell, the ancients quickly recognised the importance of good hygiene.

Of course, this isn't to say that the ancient Greeks had any sense of the role of germs in the spread of disease. Although they seem to have speculated on just about everything else, germ theory was not part of their intellectual world. However, the ancients did bequeath to their successors the idea that airborne fumes and waterborne poisons, both strongly associated with dirt and decay, tend to produce sickness. The Romans in particular invested heavily in programmes of public health. Thousands of miles of aqueducts stood as towering monuments to their belief in the relationship between filth, polluted water and disease.

Thanks to the survival of hundreds of Inquisition reports, we can even find evidence of hygienic practices among the medieval peasantry. Reports relating to the heretical peasants of Montaillou, a village in the mountains of Southern France, suggest that in some respects people lived utterly insanitary lives. Virtually no one, rich or poor, ever had a bath; in fact, nearby springs were mostly left to the lepers. Nevertheless, a raft of conventions and taboos did ensure that any part of the body involved in preparing, blessing or consuming food, the hands, face and mouth in particular, were kept relatively clean. A generalised fear of sickness as a result of swallowing foul substances led Montaillou's peasants to wash their hands before meals and vigorously scrub their faces with coarse cloths.

Poisonous Miasmas

Not until the mid-seventeenth century, however, did ideas about dirt and disease become at all precise.

Then, several British doctors and scientists, among them Robert Boyle and Thomas Sydenham, began to argue that the air contains minute, inorganic particles, often emanating from the ground, that can induce illness. By the mid-eighteenth century, this particulate, or miasmatic, theory of disease had become extremely popular. And, with the contemporary growth of large, insanitary towns and devastating epidemic diseases, hygiene began to be taken ever more seriously.

Within just a few generations, in fact, people acquired a much more refined sense of smell. The eighteenth-century urban poor were still forced to live in hastily constructed, disgustingly filthy and insanitary tenements. But awareness was already spreading that the 'pestilential human rookeries' of the larger European cities were incubators for epidemic disease. The first wave of public health campaigners drew attention to what they dubbed 'arsenals of death', the tanneries, refuse dumps and swamps in built-up areas that choked streets and dwellings with filth and foul odours. And, as the idea of the poisonous miasma became a staple of medical thought, new laws forbade the burial of the dead within city walls and, by the 1750s, steps were being taken to cover over many of the largest city sewers.

The Quaker cloth merchant John Bellers, one of the earliest public health advocates, complained most eloquently of the need to clean London's streets and to regulate where dairies and abattoirs were located.

A smooth political operator, Bellers knew how to appeal to the powers that be. 'Every Able Industrious Labourer, that is capable to have Children, who so Untimely Dies, may be accounted Two Hundred Pound Loss to the *Kingdom*', he wrote. Likewise, the philanthropist John Howard toured the prisons and hospitals of Europe during the 1770s and 1780s and returned to his native land convinced that the fevers endemic to both types of institution could be prevented through the adoption of simple hygiene precautions.

For the most part, these exhortations fell on deaf ears. Governments lacked the capital and industry the incentive to improve urban housing and sewerage systems. The most serious abuses were sometimes ameliorated, but as more people began to pour into towns unfitted for the rapid influx, things only got worse. The American doctor Benjamin Rush's 1793 advice in the event of epidemics remained the most sensible course of action for some time to come: 'fly from it!'

Yet, where eighteenth-century states did exercise control, the new 'gospel of cleanliness' often enjoyed a warm reception. The British navy, the most meritocratic and innovative of the armed services, enforced strict hygiene regulations throughout its expanding fleet. Associated with insubordination no less than illness, filth was kept at bay through regular scrubbing, whitewashing, fumigation and the use of antiseptics such as vinegar. This, together with the introduction of lime juice as a preventive against scurvy, meant that James Lind's remark that armed forces lost 'more of their men by sickness, than by the sword' ceased to apply to the Royal Navy of the late 1700s. Far more sailors were now living long enough to die of bullet wounds, drowning or venereal disease.

Despite the practical weakness of the eighteenth-century hygiene movement, by the early 1800s the notion that many illnesses were caused by inorganic airborne miasmas was thoroughly established in medical and naval circles. Only a few decades later it would feed directly into the creation of the germ theory of disease.

· CHAPTER 3 ·

CONTAGIOUS EFFLUVIA

If the notion of infection is very old, the closely related idea that disease can be spread from person to person is much less ancient. Clearly defined concepts of contagion probably only arose in Christendom in response to the horrific outbreaks of bubonic plague that scythed through her peoples during the Middle Ages. The sudden and devastating impact of the plague generated a multitude of rival explanations. Although divine displeasure, strange celestial movements and freak weather conditions were leading contenders, the nature of the disease forced many into the realisation that it also spreads from individual to individual.

Renewed bouts of plague over the following centuries only strengthened this impression. And, as evidence for the plague's contagiousness accumulated, state authorities took the necessary steps. Plague victims were isolated and their houses boarded up. Then, starting in Italy, the practice of quarantining ships' crews spread rapidly to all the major ports of Europe. The realisation that plague is contagious also inspired less benign practices. In 1347, the Tartars laying siege to the city of Caffa on the Black Sea catapulted the corpses of bubonic plague victims over the city walls in order to start an epidemic within. It was the start of the long history of biological warfare.

Contagionism gained further momentum with the introduction of syphilis from the New World. That this new and deadly illness spread by sexual contact was quickly demonstrated. Yet, as sufferers became more and more stigmatised, it came to be believed that syphilis could also be caught from the breath, clothes, utensils and even the breast-milk of sufferers. Mixed up with vague notions of 'moral contagion', respectable and God-fearing members of society learned to give the syphilitic the widest of berths.

But the definitive example of a contagious illness was smallpox. Arriving in Europe with the returning Crusaders, it eventually provided the most compelling evidence of person-to-person transmission. An appalling and often deadly sickness, famously described by Thomas Babington Macaulay as responsible for 'turning the babe into a changeling at which the mother shuddered, and making the eyes and cheeks of the betrothed maiden objects of horror to the lover', its contagious nature had been understood in some parts of the world for centuries.

Unbeknown to its European victims, by around 1000 AD the Chinese had developed a means of inoculating people against smallpox through the inhalation of dried powders made from the crusts of smallpox pustules and scabs. By the mediaeval period, it had become common practice, especially in the Middle East, to insert smallpox scabs or lymph matter with the aid of sharp ivory blades into the skin itself. But as in most aspects of scientific, economic and cultural practice, at this time Europe lagged a long way behind.

The concept of inoculation, and with it clear evidence of the contagiousness of smallpox, didn't make its way to Western Europe until the early eighteenth century. Then the highly aristocratic Lady Mary Wortley Montagu encountered it in the bazaars of Constantinople and immediately recognised its significance. She had herself been left disfigured, and eyebrowless, by smallpox and was determined to save her children from the same misfortune. Accordingly, in 1721 her daughter became the first Briton to be inoculated against smallpox.

She was soon followed by two even more wellborn guinea pigs: children of the new Hanoverian monarchy. The Hanoverians had smallpox to thank for clearing their way to the English throne, and they were understandably loath to give those next in line the same advantage. However, before George II risked his children, he had the inoculate tested on several convicted felons. Once these precedents had been set, and as the practice of inoculation became safer, it made steady progress against a fulsome barrage of criticisms.

By the time Edward Jenner began experimenting with cowpox, there were already several smallpox hospitals in London and the provinces. Every year, thousands of people would go to these hospitals in order to acquire immunity. When they were in the very best of health and most able to fight off infection, diseased smallpox matter was inserted into their skin and they would hopefully come out on top. This may sound like suicidal folly, but accounts suggest that as few as one in 400 subjects died, though a lack of standardisation of the strength of the inoculating material did mean that many more were cruelly disfigured.

So powerful a stimulus did inoculation give to the idea of disease contagion that, just four decades after its introduction into Britain, her generals were using the knowledge to deadly effect. In 1763, Sir Jeffrey Amherst instructed Colonel Henry Bouquet to infect troublesome Native Americans with smallpox by giving them rags and blankets impregnated with the bodily fluids of smallpox victims. No quarter was to be given, Amherst thundered, in the drive to 'Extirpate this Execrable race'. Having brought the Aztec empire to its knees two centuries before, this murderous disease now tore through the ranks of their Northern neighbours. 'The Indians began to be quarrelsome,' observed the Reverend Increase Mather, 'but God ended the controversy by sending the smallpox among the Indians.'

Until the late nineteenth century, doubt remained

as to whether epidemic diseases such as cholera, typhoid, typhus and diphtheria could also pass from person to person. About plague, syphilis and smallpox, in contrast, there was no dispute. This is clear from the classification of types of disease written in 1800 by the eminent Scottish doctor William Cullen. He singled out these infections as caused by 'contagious effluvia' that spread through inter-personal contact. But Cullen's account reveals that another critical step towards arriving at the germ theory had been taken. Contagious effluvia, he observed, have a marked peculiarity: they always produce the identical disease in those to whom they spread.

Here, at last, was the idea of the specific disease with a specific cause. It had occasionally surfaced before in medical thought, especially in the seventeenth-century writings of the English doctor Thomas Sydenham; but it was now becoming a basic part of medical theory. As yet, there was no conception of an organic disease agent. Nonetheless, the idea that something was passed from victim to victim and always reproduced the same illness had begun to stabilise in the medical literature. This was a major advance from the medical worldview of William Brownrigg, for whom illness had almost everything to do with personal history and virtually nothing to do with specific disease agents.

If some diseases were now thought to be both contagious and specific, it only remained for doctors to identify the microbial agents that produced them for germ theory to emerge. This convergence of concepts would not come about for several decades more. But we now turn to another strand in the process that brought it about: the steady accumulation of evidence demonstrating both the existence of microorganisms and their role in putrefaction and decay.

· CHAPTER 4 ·

LEEUWENHOEK'S 'LITTLE ANIMALS'

The first recorded dental floss took place in September of 1683. It was performed by Anthony van Leeuwenhoek, a minor city official in Holland, who became one of the finest microscope builders of his day. Having scraped some of the 'white matter' from between his 'usually very clean' teeth, Leeuwenhoek examined it beneath his microscope. Only a few months earlier this eminent Dutchman had been the first human to see microbes. But the sheer scale and diversity of the 'small living Animals' he witnessed after this historic floss simply astonished him.

One can only speculate on what it felt like to realise that any estimates one might have made as to the amount of life on Earth were fantastically, even laughably, inaccurate. As Leeuwenhoek discovered, the world teems with an utterly extraordinary variety of microbial life. Even so, it is unlikely that Leeuwenhoek felt humbled by his discovery. He knew he had made an important breakthrough and capitalised on it in a series of letters published by the Royal Society of

London. It wasn't long before people were travelling from all over Europe to press their eyes to the oculars of his microscopes, eager to catch a glimpse of this hitherto hardly imagined world.

But Leeuwenhoek didn't attribute any role in disease causation to his 'little animals'. And why should he have done? Absolutely nothing in what he saw in September 1683 jeopardised the authority of the ancient texts from which most medical theory was drawn, and in which even the possibility of germs went uncontemplated. In 1546, there had been a slight chance of a leap forward when the Italian Girolamo Fracastoro speculated that some diseases might be caused by tiny 'germs' capable of propagating themselves in their hosts. But lacking any tangible evidence and couching his ideas in robustly theological terms, it was never likely that Fracastoro's germs would subsequently be linked to Leeuwenhoek's microbes.

Indeed, the Dutchman would have been amazed to hear that his observations would one day be found relevant to explaining human disease. To Leeuwenhoek and his contemporaries, the key scientific issue was quite different: discovering where these mysterious organisms came from. Further research had shown the presence of Leeuwenhoek's microbes wherever there were signs of putrefaction and decay. This seemed to raise three distinct possibilities: germs cause rotting, they gravitate towards where rotting is taking place or they are themselves produced during the process of rotting. The third option, known as the theory of spontaneous generation, may sound an odd idea. But many of Leeuwenhoek's contemporaries were 'vitalists', scientific thinkers who imagined, quite rationally at the time, that all life depends on a mysterious inner energy or 'vital' force. They argued that once this force pervades organic, non-living matter, a pulse of life is set into motion that sometimes gives rise to the creation of micro-organisms.

Even in Leeuwenhoek's day, however, the notion that life can start unbidden wasn't universally well received. Critics of the idea were deeply disturbed by the fundamental questions spontaneous generationism raised about the nature of life itself. Is the creation of life a power reserved to the Deity? Or is it something intrinsic to organic matter that has happened every time maggots appear in cheese, wine sours, food begins to stink or mice emerge from beneath piles of rags? And, if the Creator is not involved, where does this leave the Genesis claim of a single creative burst culminating in the appearance of Adam and Eve?

These were exceedingly big issues. Anxious theologians throughout Europe complained that the theory of spontaneous generation threatened to reduce life to a mechanical property of organic matter and thereby render God redundant. Nor, they recognised, would only God's position be in jeopardy. Were His power seen to be diminished, or even extinguished, this would inevitably humble the monarchs of Christendom who supposedly ruled in His name. Small wonder that within years of Leeuwenhoek's discovery a vigorous debate was under way.

Hot Air and Gravy

So politically charged was this controversy that few scientists entered the arena without a very clear idea of what they wanted to find. Among the early protagonists was the Italian Francesco Redi. In 1688, seeking to refute the popular idea that maggots were produced by rotting meat, he placed a layer of gauze over a lump of beef and exposed it to the air. To the surprise of many, not a single maggot appeared. Redi compellingly argued that maggots could only hatch from fly eggs, none of which had been laid in his beef because of the protective layer of gauze. In the following years, talk of spontaneously generated insects and recipes for making frogs and mice largely abated. But just a few decades later, with the invention of much better microscopes revealing more and more micro-organisms, the debate flared up once more. Now it focused solely on the origin of microbes; this time, however, there was to be no single, decisive experiment.

The new generation of experimenters did at least agree on one thing: how it ought to be possible to settle the controversy. The basic techniques for investigating the origins of germs were quickly established and would hardly change in over a century. Organic matter was first heated until the existing germs could be assumed dead. Then the container was hermetically sealed so that airborne germs were prevented from settling on the sterilised mixture. Next, the experimenter waited with all the patience he could muster. Finally, after a few days or weeks, some of the experimental material was removed and examined under a microscope for evidence of life.

Yet, despite this general agreement over correct procedure, getting to the truth was far from straightforward. First, there was the unavoidable impossibility of proving a negative: 1,000 experiments providing no evidence of spontaneous generation do not eliminate the possibility of the 1,001st showing the reverse. Second, nearly all scientific experiments are open to multiple interpretations.

In 1748, the British priest-scientist John Turbeville Needham performed the standard experiment with a 'Quantity of Mutton-Gravy hot from the fire'. Unfortunately, although he didn't realise it, Father Needham's apparatus and experimental conditions were a long way short of sterile. In consequence, a few days later his sealed 'Phial of gravy swarm'd with Life, and microscopical Animals of most Dimensions'. Committed to the idea of spontaneous generation, Needham declared that 'little animals' are continually being created by a 'vegetative force' that is unleashed whenever organic materials undergo decomposition.

Because he was a priest, Needham was acutely sensitive to the charge of heresy. So, in an attempt to reconcile his religious views with his more secular scientific inclinations, he argued that the Creator had personally laid down the laws by which micro-organisms come into existence. Most of Needham's contemporaries saw this as a mere fudge. And spontaneous generation preserved its association with political and religious heterodoxy. But regardless of how he interpreted his results, Needham was convinced that he had proved that spontaneous generation really occurs.

Father Needham's time at the top was short-lived. Fifteen years later the experiment was re-run by the Italian philosopher and priest Lazzaro Spallanzani. A highly original man, perhaps the first to propose the freezing of sperm for later use, his experimental method was more sophisticated than Needham's. He sealed his phials by melting the glass of the necks before boiling the contents and leaving them to stand. When these phials were boiled for a long time, Spallanzani claimed, microbes never appeared upon cooling and the contents never underwent decay. He cleverly went on to repeat his experiment using many different forms of seal. In doing so, Spallanzani demonstrated a very close correspondence between the quality of the seal and the number of microbes that appeared in the phial several days after boiling. In showing that 'the number of animalcula developed, is proportioned to the communication with the external air', he left Needham seriously bloodied.

The Briton, however, was unyielding. There were

still enough shortcomings in Spallanzani's method to allow spontaneous generationists room to wriggle free. And, a century later, even Louis Pasteur conceded that Needham 'could not, in all justice, abandon his doctrine because of Spallanzani's experiments'. In consequence, the debate lurched on well into the following century. Finally vanquishing spontaneous generationism and shaking off the humoural world-view required the emergence of a much more rigorously scientific approach to medicine. Ironically, it was as a direct consequence of the bloody convulsions of the French Revolution that just such an ethos came into being.

· CHAPTER 5 ·

REVOLUTION IN PARIS

Hospitals and brothels for long had much in common, for at both the patron risked a combination of contagious disease and social stigma. This is because, up until the mid-1800s, most hospitals were insalubrious receptacles for those who wouldn't be there had they anywhere else to go. Where beds often held more than three people and contagious patients shared wards with the chronically ill, it is hardly surprising that for most the hospital was a brief stop-off on the way to the morgue. The rudimentary quality of medical treatment hardly helped matters. A few doctors served in a philanthropic capacity, but the vast majority had private practices and treated their patients at their own homes. In short, virtually nobody wished to be associated with hospitals.

During the progressive years of the Enlightenment, some attempts at improvement were made. A wave of hospital building had washed across Britain, Austria, France and Germany. But in the 1790s hospitals were still sinks of human suffering and degradation. 'No bugs in the beds' was the kindest thing Britain's John Howard could say of the Leeds Infirmary. 'At the hearing of that dreadful word,' wrote the British surgeon John Bell of the Parisian hospital the Hôtel-Dieu, 'the patients gave themselves up for lost'. Perhaps unsurprisingly, 'No more indigents, no more hospitals' was one of the slogans of the city's revolutionaries of 1789.

In the event, hospitals didn't go the same way as the monarchy and the Bastille following the turmoil of political revolution. Instead, the French state overhauled the way in which they were run. None of these reforms was entirely novel, but the scale of the changes and the energy with which they were carried out did lead to a genuine revolution in medical practice. The reforms instituted in Paris became the model that the rest of the civilised world followed, and this paved the way for an extraordinary shift in the doctor's understanding of infectious disease.

The Decline of Deference

In the immediate aftermath of the revolution, huge metropolitan hospitals were established, each with a permanent staff of qualified doctors. Partly because many of their richest patients had gone to the guillotine, these doctors now had the time to give students practical instruction on their wards. And, within a few years, students from all over France, then Europe and America, were flocking to Paris' hospitals, eager to benefit from the unrivalled opportunities for medical instruction. Like George Eliot's fictional Dr Tertius Lydgate, they returned to their native shores enthused with a new professionalism.

Rather than spending years learning Latin and Greek, poring over ancient texts, and hardly seeing a patient until fully qualified, these trainee physicians found themselves learning at the bedside or in the autopsy theatre. 'Read little, see much, do much' was the motto of the reformist Antoine Fourcroy. And what students now saw with their own eyes was all too often at variance with what the ancients had taught. In this atmosphere of critical enquiry, traditional dogmas began to lose their aura of infallibility.

Major structural changes were also at work here. As we have seen, there were social as well as medical reasons why humoural theories of disease had such a long shelf-life. For wealthy eighteenth-century patients liked their doctors to be their intellectual equals, but no more. This was soon to change. Prizing many of the best physicians away from their clients, putting them on the state payroll and letting them loose on thousands of poor hospital patients revolutionised the doctor–patient relationship. Instead of kow-towing to a few rich clients, the élite doctor became the imperious master of the hospital ward. He thereby gained complete intellectual freedom. And, as the patient's grip on what he could say and do declined, medical ideas started to become much more sophisticated.

This decline of traditional deference meant that other social niceties, which had for long hindered medical progress, quickly disappeared. Back in 1737, Queen Caroline, wife of George II, had begun complaining of severe abdominal pains. Unfortunately, an acute sense of social propriety prevented her doctors from undertaking a proper physical examination. Instead, the Queen was relentlessly bled, purged and cupped. Only in response to a Royal command did the physicians eventually examine the Queen's abdomen, at which point they found she had a strangulated umbilical hernia and was rapidly dying of gangrene. Deference, delaying as it did proper surgical intervention, made doubly sure that the Queen died. In striking contrast, in Paris of the early 1800s, patients were stripped, felt, prodded and examined as closely as the doctor felt necessary for effective diagnosis.

This scientific turn had another stimulus. If the new élite doctors took liberties with the living, they had *carte blanche* with the dead. The dissection of human bodies now became a central part of medical training and research. Fresh bodies had always been a rare treat for the medical student. Surgeons had traditionally loitered around sites of public execution in the hope of purchasing the bodies of the condemned for anatomy instruction. Others were forced to pay 'resurrection men' to plunder graveyards. All this changed in the late 1700s, as patients who could not pay for their hospital care willingly or unwillingly left their bodies to medical science.

Doctors who now had the luxury of cutting open hundreds of cadavers a year could begin studying disease at close quarters. Both literally and figuratively, their investigation of disease was going deeper than ever. In the past, they had classified illnesses according to symptoms alone; hence their fixation on bodily fluids, their balance, retention and excretion. But autopsy allowed doctors to see past these superficial signs. 'Obscurity will soon disappear', enthused the eminent Marie Xavier Bichat. And so, in many respects, it did. Their exploration of internal diseases started to undermine the entire edifice of humoural medicine.

As doctors sliced, sectioned and minutely inspected the organs of the human body, they began distinguishing between illnesses with new precision. Pierre-Fidèle Bretonneau, for instance, discovered that fevers with very similar symptoms can be associated with quite different kinds of damage inside the patient. Following Bretonneau, many other doctors found clear correlations between certain diseases and lesions located in particular parts of the body. Most maladies, they noticed, affect individual organs and tissues. In consequence, the idea of 'peccant humours' moving from one part of the body to another, giving rise to a different illness in each, soon made far less sense. Just as importantly, the idea of disease specificity received a powerful boost.

This gradual realisation that each disease is a distinct kind of infection also owed much to the environment in which these hospital doctors were working. Seeing hundreds of patients every year in their enormous wards drew the doctors' attention to the common features of disease. Rather than sickness being a matter of the individual constitution, they increasingly recognised that most patients actually suffer from a narrow range of specific ailments. What mattered to doctors now was what patients had in common, not the slight idiosyncrasies that made them different. Medicine à la Dr Brownrigg was fast falling out of favour. Within a few decades, humoural theory would be left in ruins and in its place briefly triumphed this new concept of disease as involving damage to specific parts of the body. It was a way of thinking about illness that would take doctors a few steps closer to accepting the germ theory of disease.

The Rise of the Laboratory

This growing appreciation of the specific nature of disease soon raised a question that could only be answered with the advent of experimental medicine: did these specific illnesses also have specific causes? Within a few decades, many of the new metropolitan hospitals had become centres for rigorous experimentation in which this kind of question could be addressed. Plenty of hospital doctors now had the time, money and resources to put medical theories to the test and utilise ever-improving microscope technology in investigating disease. In parallel, vivisection became a staple of medical research. The Frenchman Claude Bernard's cruel but compelling experiments on dogs permitted real advances in scientific understanding of the function of the liver and other bodily organs.

In time, government and industry also came to appreciate the value of medical research. Realising that improvements in medical science were essential for the health of their industrial workforces and as a basis for national prestige, they began allocating substantial funds to scientific and medical research. The state-funded laboratory was by far the most important product of this change of heart. Where France had been the pioneer of hospital medicine, by the late 1830s the Germans had assumed the lead by establishing an exceptional collection of research laboratories. Presided over by Johannes Müller, Jacob Henle and Justus von Liebig, they soon became the envy of the world. As a result, this laboratory revolution eventually swept through North America and the remainder of Europe as well.

Thus, in large part due to the changes wrought in the wake of the French Revolution, by the 1850s the medical profession had travelled a long way in a remarkably short period of time. Compared to Brownrigg's world of haphazard research, in which doctors seldom looked beyond their patient's symptoms, hospital medicine could now boast a culture of scientific rigour. Like its political midwife of 1789, however, perhaps the most important effects of the Paris revolution were social. The medical scientist had emerged as a prestigious figure from whom important practical discoveries were eagerly expected. For the great majority of medical men, there would be far less bowing and scraping to a few wealthy clients. They were intellectually independent and had at last become the custodians of specialised medical knowledge. In short, this was an environment in which new discoveries were always likely. And radical ideas, if pushed hard enough, could ultimately win through.

· CHAPTER 6 ·

DIRT, DISEASE AND DECAY

It wasn't long before the scientific ethos derived from Parisian medicine translated into scientific advance. Four decades after the Italian priest Spallanzani had sought to embarrass Father Needham, his experiments were repeated by the outstanding Austrian scientist, Theodore Schwann. Schwann was eager to disprove an objection levelled by the advocates of spontaneous generation against Spallanzani's methods. They had asserted that germs failed to arise in Spallanzani's phials solely because there wasn't enough oxygen for spontaneous generation to occur. Given the prevailing level of scientific knowledge, this was a perfectly legitimate criticism of Spallanzani's results. But Schwann responded to it with the same flair for scientific research that later made him famous for showing that cells are the basic units of all life.

In 1845, he set up an experiment in which heat-sterilised beef broth was exposed to a continuous supply of fresh air. The organic mixture was to have all the oxygen it could possibly require, but on its way to the phial this air was first heated to 100°C. No doubt to Schwann's delight, his heat-sterilised phials were still free from germs when examined several days later. Nor was there the slightest evidence of putrefaction.

But was the case at long last closed? These experiments certainly seemed to prove that airborne microbes cause decay. The debate, however, continued. And this wasn't simply a case of the proponents of spontaneous generation refusing to face facts. Good science requires scientists to seek relentlessly to disprove one another's results. And Spallanzani and Schwann had left plenty of questions unanswered.

For a start, having tried to replicate Schwann's results, many scientists found it hard to swallow his claim that 'this experiment was repeated many times, and in all cases there was no putrefaction even after many weeks'. Schwann's is a hard experiment to perform and one suspects that his claims of success were somewhat exaggerated. Today we realise that his critics' problems stemmed from the difficulty of totally excluding germs from the experiment. But during the 1830s and 1840s, how could one tell if microbes sometimes appeared after boiling because the apparatus was faulty *or* because spontaneous generation had actually occurred?

Apparently undaunted, Schwann next set out to refute another of Needham's defences. Repeated boiling, the English priest had argued, sometimes destroys the capacity for organic matter to generate new life. 'From the way he has treated and tortured his vegetable infusions,' he wrote of Spallanzani, 'it is obvious that he has ... much weakened, and maybe even destroyed, the vegetative force of the infused substances.' In response, Schwann tried to demonstrate that the kind of putrefaction that an organic mixture undergoes, i.e. the sorts of chemicals produced, is entirely dependent on the airborne matter that falls upon it. Using a poison that killed both bacteria and moulds, Schwann showed that his beef broth always failed to decay. But a poison that singled out bacteria alone allowed mould to grow quite happily on the surface of the broth. In neither case could evidence of spontaneous generation be found.

Schwann's argument was clever, but hardly conclusive. Wasn't it possible that the poisons he used had destroyed the 'vegetative forces' just as surely as boiling? Nonetheless, by the 1850s Schwann had won many converts. It was now entirely plausible to think of germs as the cause of decay. And even if the causal relationship between micro-organisms and putrefaction remained unproven, more and more people were beginning to think that if germs can cause meat to rot and wine to turn sour, they might also produce dangerous and deadly reactions inside the human body. It was just this imaginative leap that later provided both Louis Pasteur and Joseph Lister with the inspiration to start thinking about the role of germs in causing disease.

Yet, science rarely proceeds on one steady tack. Nature is generally far too complex for a handful of experiments to yield definitive answers. And, despite Schwann's ingenuity, in the following years it began to look as if the vessel he had piloted so skilfully was going to be driven back onto the rocks.

Liebig and the Public's Health

In 1839, the brilliant German chemist Justus von Liebig stepped into the fray. Liebig announced his belief that fermentation and rotting are strictly chemical, and not biological, processes. In saying this, he was not supporting the idea of spontaneous generation. He simply believed that microbes are no more than secondary colonisers of putrefacting substances. Microbes may be present, Liebig stated, but they are not necessary for decomposition to occur. He went even further. The claim that microbes cause putrefaction is as ridiculous as 'saying that the causes of the decay of wood ... arise from the plants which use these decaying products as nutrients'.

Literally never one to pull his punches, as a young chemistry professor involved in radical politics, Liebig had been thrown into jail for striking the hat from the head of a policeman. Since then, he had done more than any other individual to make chemistry a prestigious scientific discipline, and his forceful personality dominated the field for decades. And out of deference to Liebig's tremendous stature, many of those who had earlier supported Schwann now started to change sides.

But Liebig's status was not the only factor. His argument was also entirely credible: just because germs are always found at the site of decay is not to say that they are essential for it to take place. Microbes seek out rotting substances, he insisted, rather like vultures gravitate to carrion on the African savannah. They are not responsible for the deaths, but they eagerly exploit them. Clearly, then, neither the big guns nor the best arguments were all on the side later proven to be right. So cocksure was Liebig of his explanation that in 1839 he co-authored a skit entitled 'The Riddle of Vinous Fermentation Solved'. It ridiculed Schwann and his allies by depicting a micro-organism eating sugar, then ejecting alcohol through its anus and carbon dioxide through its genitals.

Yet an amusing cartoon wasn't Liebig's only positive contribution to this story. Often cast in histories of the germ theory as an arrogant reactionary, his influence was actually more complex. Sure, he put the dampers on attempts to find disease-causing microbes. But his ideas on fermentation were also a key impetus behind the Victorian obsession with the role of airborne particles in the origins of epidemic disease.

Liebig's claim that putrefaction is really the result of chemical agents that cause organic material to decompose, greatly enhanced interest in the notion of miasmas, first clearly articulated by Boyle and Sydenham a century earlier. Hundreds of doctors in both Europe and America now took up this idea. And, inspired by the eminent German, it became fashionable to see sickness as the result of reactions produced inside the body by dangerous chemical miasmas. These imagined minute, toxic particles differed from the later notion of germs only in that they were not considered to be alive.

In Britain, the doctor and statistician William Farr took Liebig's theory of fermentation a few steps further in seeking to narrow the gap between the chemical and biological realms. In an 1842 classification, he introduced the notion of 'zymotic' diseases, a label covering most infectious and contagious illnesses. These, Farr believed, were caused by organic particles, often likened to the pollen of flowers, that were ingested or inhaled and then started a process of chemical decay inside the body. Farr's organic particles were not thought to be the microbes Leeuwenhoek had seen (he would remain a staunch critic of the germ theory throughout the 1850s). But Farr's ideas were still a sizeable step in the right direction; for in buying into miasma theory, his followers accepted that very small entities could lay low, or destroy, the very largest and most complex of life-forms.

The theories of Liebig and Farr also helped to inspire the large-scale public health campaigns of Edwin Chadwick and John Simon in Britain, Lemuel

Shattuck in America, Rudolf Virchow in Germany and Louis Villermé in France. Although pitched against powerful middle classes that venerated private enterprise and distrusted central government, reformers succeeded in realising many of Bellers' and Howard's eighteenth-century dreams. Throughout Europe, and then America, official reports detailed the insanitary horrors of urban slums and rural shanty towns, bringing home to the middle classes the tight relationship that held between filth and sickness. As the influential classes were made to see that outbreaks of disease starting in working-class tenements could spread to their own more salubrious environs, governments were pressured into constructing thousands of miles of sewerage tunnels to carry foul water away and complementary systems to bring in fresh supplies. The tremendous impact this engineering work had on mortality rates was taken to be a powerful vindication of miasma theory.

Hospitalism and Hygiene

It is one of the greatest ironies of medical history that, unintentionally, hospitals provided some of the strongest evidence for the infectiousness of many forms of disease. The Parisian model of hospital medicine had wrenched the élite physician away from his private patients and revolutionised medical education and research. But the rapid growth of urban hospitals served only to make them even less healthy than before. The term 'hospitalism', used to describe hospital epidemics, was coined by the Scottish surgeon James Young Simpson. He famously remarked that a patient 'laid on an operating-table in one of our surgical hospitals is exposed to more chances of death than the English soldier in the field of Waterloo'. Yet, had hospitals been less insanitary, doctors would have been far slower to appreciate the infectiousness of disease.

During the 1850s, Florence Nightingale, already raised to the status of a secular saint for her labours in the hospitals of Turkey and the Crimea, rose to the fore in an evangelical campaign to combat hospitalism. With seemingly boundless energy, she devoted herself to ensuring that hospitals were well ventilated and free from the piles of refuse, discarded, bloodied swabs and filthy bedclothes typical of the pre-1850 period. And Nightingale wasn't alone. Already under way by the time she became involved was a concerted drive to isolate surgical patients, regularly white-wash wards and scrub walls with disinfectants such as carbolic acid. Although now largely (and undeservedly) forgotten, George Callender, head surgeon at London's St Bartholomew's hospital, spearheaded this movement. Between 1847 and 1867, Bart's mortality rate from wound infection fell from roughly 35 per cent to a mere 10 per cent. By 1877, it had fallen still further to just 2 per cent.

In view of these achievements, it may seem remarkable that most surgeons and many public health campaigners of the 1850s didn't believe that sickness could be passed from one person to another. Epidemics, they argued, always occur in the midst of filth and poor ventilation, be it in dirty hospitals or insanitary urban courtyards. So isn't it enough, they asked, to implicate toxic miasmas emanating from squalid surroundings? 'There is no proof, such as would be admitted in any scientific inquiry, that there is any such thing as "contagion"', Nightingale asserted in 1859. To her and her colleagues, what should be done was obvious: disinfect wards, clean the streets, lay networks of sewers and prohibit the dumping of animal carcasses in built-up areas and epidemic disease would recede like mist from the rays of the sun.

There was, however, more to this attitude than simple confidence that the miasma theory explained everything. Many establishment figures were opposed to the idea of contagion because they had an instinctive aversion to repressive quarantine measures that hindered trade and compromised individual liberty. But opposition also came from within the medical profession itself. The doctor's approach to childbed, or puerperal, fever sheds light on why this was so and, more importantly, how it came to be challenged.

The Angels of Death

Perhaps because the death of a mother soon after childbirth is so inherently tragic, the lethal consequences of childbed fever always loomed larger to the medical profession than many deadlier diseases. Yet this concern did little to increase the doctor's understanding of how the infection spread. From the end of the eighteenth century, a few did begin touting the correct idea that it is contagious; but not until the 1870s was the case widely accepted. Its ultimate success proved an uphill struggle largely because both doctors and midwives had ulterior motives for refuting the contagionist's arguments.

The first doctor brave or foolhardy enough to champion the contagionist argument was Scotland's Alexander Gordon. Between December 1789 and March 1792, his neighbourhood was hit hard by an epidemic of childbed fever. Gordon immediately set about documenting the cases and he soon found that they all had something in common: in every instance, the doctor or midwife present at a fatal delivery had previously attended to another 'patient labouring under the disease'. Childbed fever, Gordon concluded, is contagious and 'is transmitted from one case to another by doctors and midwives'.

He was right, but this was hardly the sort of thing his colleagues wanted to hear. It was fine not always being able to cure one's patients, but no medical professional welcomed being dubbed, in effect, an angel of death. Gordon made things even worse by naming in his treatise the midwives associated with the highest number of deaths. Small wonder that they teamed up and did all they could to discredit him as a purveyor of dangerous new ideas. Gordon was driven out of Aberdeen and, after a brief stint in the navy, he retired, broken and disappointed, to his brother's farm.

In comparison, the American poet and physician, Oliver Wendell Holmes, got off rather lightly. Yet he too provoked a furious response when, as a newly qualified doctor, he arrived at the same conclusion as Gordon. Holmes noted that there were several striking cases of individual doctors losing dozens of patients to childbed fever, whereas others in the same locality had lost none at all. Clearly, then, the incidence of the disease couldn't be explained in the miasmatist terms favoured by the vast majority of his profession; diffuse clouds of poisonous particles can't single out individual practitioners. Dr Rutter of Philadelphia, for example, saw mothers develop childbed fever after every delivery he attended over the course of a year. This was obviously no coincidence.

Rutter himself had realised that the cause must be something that he carried on his body or his clothing. So he had his head and face shaved, washed thoroughly, changed all his clothes and instruments, and quarantined himself for several weeks. Still, having begun practising again, mothers under his care kept dying of childbed fever. Soon after, Rutter left Philadelphia in a hurry, his reputation damaged beyond repair.

In 1855, Holmes cited this and other evidence to the effect that 'puerperal fever is so far contagious as to be frequently carried from patient to patient by physicians and nurses'. He made no mention of living germs. But he did implore doctors and midwives to avoid the post mortems of women who had died of childbed fever and to cease attending deliveries if more than three women in a row died while under their care.

The critical response to Holmes' book was swift and savage. The eminent Professors Charles Meigs and Hugh Lodge of Philadelphia led the charge. Incensed at the self-assurance of this obscure young physician, Meigs declared Holmes' essay to be the 'maunderings of a sophomore'. Lodge told his students that it was simply unthinkable that physicians were to blame for the spread of childbed fever. Nevertheless, from about the 1840s, many American doctors did begin to hedge their bets and wash thoroughly before delivering babies.

In Britain too, the idea of contagion now made real headway. During the 1840s, James Young Simpson became thoroughly convinced that doctors and midwives spread childbed fever. Among the first to suggest that the germs of disease travel upon their fingers, he memorably likened these to 'the ivory points formerly used by some of the early vaccinators'. Simpson revealed that childbirth often causes abrasions to the vagina, leaving the uterus raw and the 'mouths of numerous arteries and veins' exposed. This, he went on, provides any number of points at which deadly substances can enter the mother's body.

As Simpson was developing these ideas, another

doctor, born in Hungary but working in Vienna, came to the same conclusion. Ignaz Semmelweis was horrified by the rates of childbed fever in the Vienna General Hospital and set himself the task of establishing their cause. Fortunately for Semmelweis, the presence of two maternity wards with very different death rates afforded excellent opportunities for comparative study. In the first ward, the mortality rate was an appalling 29 per cent. In the second, it was only 3 per cent. Semmelweis was struck by this disparity and, before long, he had found its cause. The births in the first ward were handled by medical students who often went straight from the autopsy room to the maternity ward. But only midwifery students worked in the second ward and their hands were free of cadaverous matter. Semmelweis realised that the doctors were routinely, if unknowingly, transferring lethal material from the dead to the living in a wretched, ongoing cycle.

As if fate were trying to ensure he drew the correct conclusion, he soon after witnessed the death of one of his own professors as a result of accidentally cutting his finger during an autopsy. The professor's symptoms were precisely like those of the mothers dying by the dozen in Maternity Ward One. So, in May 1847, Semmelweis took the dramatic step of ordering his staff to wash their hands in chlorinated water before deliveries. Deaths from childbed fever were suddenly and spectacularly reduced.

Yet, rather than achieving celebrity status, only a

decade later Semmelweis was dying in a low-grade insane asylum, abandoned by his wife, and forgotten or reviled by most of his ex-colleagues. Although many accounts of what went wrong mirror the treatment meted out to Gordon, the real reasons for Semmelweis' tragic fall are more complex. Certainly, his ideas irritated some hospital superiors who refused to admit that they had been the unintentional cause of hundreds of deaths. But the classic story in which Semmelweis is instantly ostracised for his views by self-interested doctors is almost pure myth. In fact, he rapidly won many loyal supporters who were at first happy to travel around Europe spreading news of his achievements. These dedicated followers only began to fall away once Semmelweis underwent a major personality change, which coincided with him being refused a post he had coveted at the Vienna hospital. In the months that followed, his hitherto often eccentric behaviour shifted into the realms of psychosis.

Violent mood swings, extreme paranoia and sexual deviancy now rendered him the worst imaginable advocate for a radical theory. Semmelweis scandalised the medical establishment by writing of one of his critics, 'I denounce you before God as a murderer', and he suggested that another 'take some semesters in logic'. Leaving Vienna without making any effort to say goodbye to his friends, Semmelweis alienated most of his remaining allies. This was a disaster. Getting original scientific ideas accepted requires exceptional social skills. But Semmelweis' mental state denied him the capacity to act effectively. Personal misfortune, and not simply the arch-conservatism of his peers, prevented his hygienic practices from being taken up more widely.

Nevertheless, by the time of his incarceration, Semmelweis' views, as well as those of Gordon, Holmes and Simpson, had been circulated extensively. And, in the following decades, a growing number of doctors were prepared to accept the contagiousness of childbed fever. Changing attitudes to this particular disease, perhaps more than any other, now prepared the ground for a much bigger breakthrough.

· CHAPTER 7 ·

THE END OF THE BEGINNING

By the middle of the nineteenth century, the stage was set for the germ revolution. Few saw it coming, but enough was known about microbes, their possible role in causing decay and the existence of contagious disease, for the germ theory to seem at least worth investigating. Ideas of poisonous miasmas, whether organic or inorganic, had also prepared the minds of many doctors for the possibility that tiny particles might cause disease. Yet, for all this, advocates of the germ theory in 1850 thought the future looked pretty bleak.

It was in this year that John Simon, Britain's great public health reformer, addressed his colleagues and students on the ideas of a Swiss doctor called Jacob Henle, who had suggested that living micro-organisms cause many serious diseases. Simon resolutely set his face against such a possibility. Paying homage to Liebig, he expressed a firm belief that 'the phenomenon of infective diseases is essentially chemical'. This was an opinion from which few in his audience demurred.

It is tempting to explain the resistance of Simon and his myriad allies in terms of the bloody-mindedness of doctors brought up in older medical traditions. This is how the history of science used to be written: a few prophetic geniuses realise how things really work and then spend years in the wilderness, shunned by colleagues and rivals unwilling to admit their mistakes or too caught up in what they've been taught to see the truth. In reality, accounts like these are mostly romantic fictions. The scientist, like anyone else, can be stubborn, hostile to change and motivated by envy and pride. But we too often overlook just how difficult proving a new theory can be.

The history of germ theory exemplifies this point. Take, for example, the case of John Snow, doctor, epidemiologist and obstetrician to Queen Victoria. In the midst of the London cholera epidemic of 1854, he showed that deaths from the disease in one neighbourhood of Soho clustered around Broad Street's water-pump. Partly on this basis, he put forward a radical thesis: that cholera is a waterborne disease, that it is caused by a specific germ and that it is released in the stools of the infected. Today, Snow is an international hero. Celebrated in textbooks of biology, public health and epidemiology, this passionate teetotaller is even immortalised in the name of a Soho pub, 'The John Snow'.

Yet, during the 1850s, *The Lancet*, Britain's leading medical journal, accused Snow of drawing his ideas from the 'main sewer' and rejected all his articles. The

French Academy of Science also refused to publish any of Snow's papers. As science has vindicated Snow, his many opponents now look pretty silly, if not downright malign. But if we go back to the 1850s, we begin to see that there were several holes in his argument big enough to drive a hearse through.

In reality, Snow's evidence of deaths clustering around a pump in no way proved that contaminated water caused the Soho cholera epidemic. As several epidemiologists observed, his data were equally consistent with the rival claim that a pile of putrefying household waste in the region of the pump had released miasmas and poisoned the Soho residents. But even among those prepared to accept that the Soho pump water was the source of the outbreak, there were many who accused Snow of gross credulity for believing that cholera could only be caused by water infected with the faecal matter of other sufferers. Logically, these critics were making a sound point: Snow was going well beyond the available evidence in trying to extrapolate from a few cases in one outbreak to include every occasion on which the disease had struck.

Those rejecting Snow's ideas also included doctors whose entire training had been built on the belief that most diseases can be produced in a number of different ways. Snow offended these deeply ingrained sensibilities in suggesting that cholera couldn't possibly occur without the presence of a specific germ. But once again, and not least because there seemed to

Wentworth Street, Whitechapel.

Illustration 2: An image of a typical, insanitary metropolitan slum of the Victorian period and its unfortunate denizens. Reproduced from Gustave Doré and Blanchard Jerrold's *London: A Pilgrimage* (London: Grant, 1872), f. p. 124.

be no evidence that cholera germs actually existed, he was quite reasonably thought to be overreaching himself. 'Where are these little beasts?' mocked Edinburgh's John Hughes Bennett, 'Has anyone seen them yet?'

As it happens, by 1854 this accusation was no longer valid. In that year, the Italian Filippo Pacini became the first to see the cholera bacillus. But his quickly forgotten account showed only that a comma-shaped bacillus tended to be found in the stool of people with cholera but not in the stool of the healthy. This wasn't enough to prove that the bacillus was the cause of cholera. It seemed equally plausible to argue that germs were no more than secondary colonisers of rotting food or diseased tissue. 'It is still undecided whether putrefaction follows from bacteria or bacteria simply appear where there is putrefaction', remarked the distinguished German surgeon, Carl Thiersch, in 1875. Moreover, while the case against spontaneous generation remained unclosed, just showing some strange entities on a microscope slide really wasn't enough.

Even when germs were found at the site of infection, the majority of doctors were right to be sceptical of claims that they were its cause. In 1869, for example, the American doctor James H. Salisbury announced that he had identified types of fungi that caused measles, typhoid and malaria. This was soon shown to be nonsense when others noted the occurrence of these illnesses in geographical regions where the offending fungi were never found. To drive the point home, the Philadelphia doctor Horatio Wood downed a glass of water containing large quantities of Salisbury's fungi with no ill-effects.

The British doctor Lionel Beale eloquently advanced a final objection to the germ theory. In 1868 he showed that bacteria exist in the digestive and respiratory tracts of the perfectly healthy human body. We now know that, from a human perspective, there are both good and bad bacteria. But Beale saw his findings as evidence that germs pose no threat to mankind, and his view found plenty of supporters.

Clearly, then, by the 1850s the germ theorists had it all to do. There was still no compelling evidence for the role of micro-organisms in causing internal diseases. It wasn't even established that microbes cause decay outside the body, and there were plenty of observations of harmless bacteria suggesting that Leeuwenhoek's 'little animals' could not possibly produce ill health. So the major challenges to the germ theorist were, first, to vanquish spontaneous generationism and, second, to prove that specific microbes consistently produce the same diseases in their hosts.

Fortunately, medical science at mid-century was bursting with new potential. And one man more than any other embodied the new-found appetite for experiment as the royal road to new discoveries. He was the Frenchman, Louis Pasteur.

Part III

Cue, Louis Pasteur

· CHAPTER 8 ·

TWO DUELS

We left the debate about spontaneous generation at an impasse during the 1850s. The immense prestige of the chemist Justus von Liebig had led many to follow him in rejecting Spallanzani and Schwann's evidence that germs cause putrefaction. It was at this critical juncture that Louis Pasteur took the stage. His family had made their way as tanners in the sedate French town of Arbois, but their son Louis was intensely and often ruthlessly ambitious. This appetite for glory may have been whetted by his father proudly recounting stories of his heroism during the Peninsula War. But, whatever its origins, the young Pasteur was convinced he had within him the stuff of greatness. While this self-belief could sometimes make him seem unbearably conceited, by the 1850s medicine sorely needed someone with the arrogance to reassure his wife at the age of 30 that he would 'lead her to posterity'.

Liebig's Licking

Pasteur's first breakthroughs arose from his knowledge

of chemistry rather than biology. In the mid-1850s, he was a chemistry professor based in Lille. Since this was a region dominated by brewers, vintners and vinegar makers, it was only natural for him to develop a keen interest in the chemistry of fermentation.

Pasteur's research project concerned the crystalline structure of tartaric acid, a compound usually found in sour wine. His experiments using polarised light revealed that the acid's crystals can exist in two different forms that are, atomically speaking, mirror images of each other. This may not sound too exciting. Never before, however, had a chemist detected such an asymmetrical molecular arrangement: Pasteur had discovered something unquestionably novel. But was it important?

Intrigued and delighted, Pasteur set out to discover how these asymmetrical compounds are formed. He eventually arrived at the right answer: they can only be produced through the involvement of living micro-organisms. And, since tartaric acid is often formed during wine making, Pasteur came to believe that fermentation also requires the action of microbes. Thus, by a rather circuitous route, he now found himself embarked on a collision course with Justus von Liebig. It was to prove a violent encounter.

Liebig's opposition to germ theories of fermentation rested on his belief that wine, beer and vinegar are produced through an exclusively chemical process. Germs are only present in the midst of decomposing or fermenting matter, he insisted, because that's

Illustration 3: A portrait of Louis Pasteur. Etching by Champollion, 1883. Source: The Wellcome Library, London.

where they most like to be. But Liebig had the double misfortune of being both wrong and up against a clever and determined adversary. Within just a few years, the Frenchman had shown not only that fermentation relies on living micro-organisms, but that it takes one micro-organism to make fine wine, another to make vinegar, another to make beer, and still others to produce putrid or foul-smelling slops.

Pasteur's aim in 1857 was to prove that the yeast used in the process of fermentation comprises micro-organisms and not chemical compounds. Cagniard de Latour had tried to do the same thing two decades earlier and had reported seeing 'simple, transparent, spherical or slightly elongated, almost colourless' globules that looked exactly like the living cells. But Liebig had blithely cast Latour's conclusions aside as inconclusive. This made it clear to Pasteur that much more evidence was required to prove that yeast is actually alive. He needed to show that during the process of fermentation, yeast multiplies and grows just like any other living organism in the presence of appropriate nutrients.

So he began by mixing together a solution of sugar-water, yeast and ammonia and, once fermentation was concluded, he then measured the quantities of the various elements left behind. Doing so revealed that carbon from the sugar-water and nitrogen from the ammonia had been absorbed by the yeast and, he noted with excitement, used to produce new cell-like globules. Strictly proportionate to the amount of nutrients provided, the quantity of yeast had increased. It had, in short, acted just like a living organism.

Following up this experiment, Pasteur spent interminable hours 'at night using gaslight for illumination', poring over his microscope watching how new yeast globules are formed. Everything he saw suggested that he was watching living cells in the process of formation. Older globules could be identified because of their thick membranes, and from these every so often appeared what looked just like tiny buds. Gradually, these buds grew larger, developed harder membranes, and then partially separated from the 'mother' globule. The same process, repeated hundreds of times, led to the creation of long, serpentine chains of yeast cells; this looked nothing like an inorganic chemical reaction. And, combined with the virtuosity of Pasteur's earlier experiments, these easily repeated observations now convinced much of the scientific community that Liebig had got it all wrong.

In the course of these studies, Pasteur made another vital breakthrough. He was able to show that wines, beers and vinegars contain micro-organisms that vary in appearance, and he leapt to the conclusion that different germs have different effects. His critics were quick to point out that although germs might look dissimilar, they may behave in exactly the same way. Pasteur made short shrift of this claim with a superb series of experiments in which he first isolated particular forms of microbes and then sowed them upon various fluid media.

This technique soon demonstrated that the kind of microbe sown determined whether the final product was alcohol or something else. For example, whereas just adding yeast to a sugar solution always gave him alcohol, the addition of a bacterium he named *Mycoderma* produced vinegar. Further evidence for the existence of specific microbes came from his discovery of an organism that produced butyric acid but could function only in oxygen-free environments. Bacteria, these experiments suggested, aren't jacks of all trades. Rather, each performs the specific roles allocated to it by nature, and these roles alone.

Pasteur delivered the *coup de grâce* to his opponents by experimenting with a fluid close to the Gallic heart. Impressed by reports of his earlier work, the Emperor Napoleon III personally invited him to investigate the 'wine diseases' that bedevilled viniculture. Politically a fierce conservative, Pasteur happily accepted the invitation. Joined by his able lieutenant Emile Duclaux, he set up a laboratory in his home town of Arbois. Within weeks, they had shown the presence of contaminating bacteria in a batch of yeast that had produced sour wine. Isolating the bacteria and sowing them in other wine barrels caused these to turn bad as well. Over the subsequent months, Pasteur and Duclaux identified many other kinds of bacteria, some of which made wine tasteless and sparkling, while others reduced it to rancid slops.

Thanks to their labours, by the 1860s wine production had become a far less risky venture. Pasteur and Duclaux had shown how vintners could use the microscope to identify which barrels were contaminated. Their contents could then be poured down the drain instead of wasting yet more time and money. Better still, Pasteur revealed that controlled heat could be used as a means of destroying harmful bacteria in the finished wine. 'Pasteurisation', made yet more famous in the processing of milk, was born.

With this string of achievements behind him, Pasteur's ascent to national celebrity was off to a flying start. Liebig watched his progress with increasing alarm. In 1871 he published a riposte that had been 10 years in the writing. But the outcome of this battle was the reverse of the crushing military defeat the French had just suffered at the hands of the Germans. Pasteur now challenged Liebig to boil a barrel of vinegar during fermentation. According to Liebig's chemical theory, doing so would not impede the production of vinegar in any way. But if micro-organisms are required, then fermentation would stop due to the destructive effects of the heat. The French Academy of Science sniffed victory in the air and immediately wrote to Liebig volunteering to fund the trial. Sensing defeat, Liebig failed even to respond.

The Flight of Felix Pouchet

Pasteur's involvement in fermentation research

inevitably drew him into the interminable debate over spontaneous generation. When, in 1858, his countryman Felix Pouchet threw down the gauntlet and announced that he could prove that spontaneous generation occurs, Pasteur was the obvious person to take up the challenge. Pouchet, a distinguished scientist in his own right, claimed that 'boiled hay-infusions', kept in an air-tight trough of mercury, showed clear signs of microbial life after they had been left for several days or weeks. Upon leaving Lille and returning to Paris, Pasteur set about disproving this. Only a few years later, in 1864, he exhilarated the cream of French society gathered at the Sorbonne with a description of how he had consigned spontaneous generation to the graveyard of bogus scientific ideas.

Pasteur recounted an epic struggle between himself and Pouchet, involving a series of charges and counter-charges as the battleground moved from his Parisian laboratory to the mountains of the Pyrenees and to the Mer de Glace, high in the French Alps. As he closed his speech, Pasteur could claim a historic victory and the flight of a vanquished and humbled Pouchet to his native Rouen. The French scientific establishment breathed a sigh of relief that this chapter in the history of science was finally closed.

The experiments that Pasteur recounted at the Sorbonne were similar in style to those of Spallanzani and Schwann. He did, though, contribute some brilliant innovations. For instance, he placed boiled sugar-water in swan-necked flasks, the necks of which

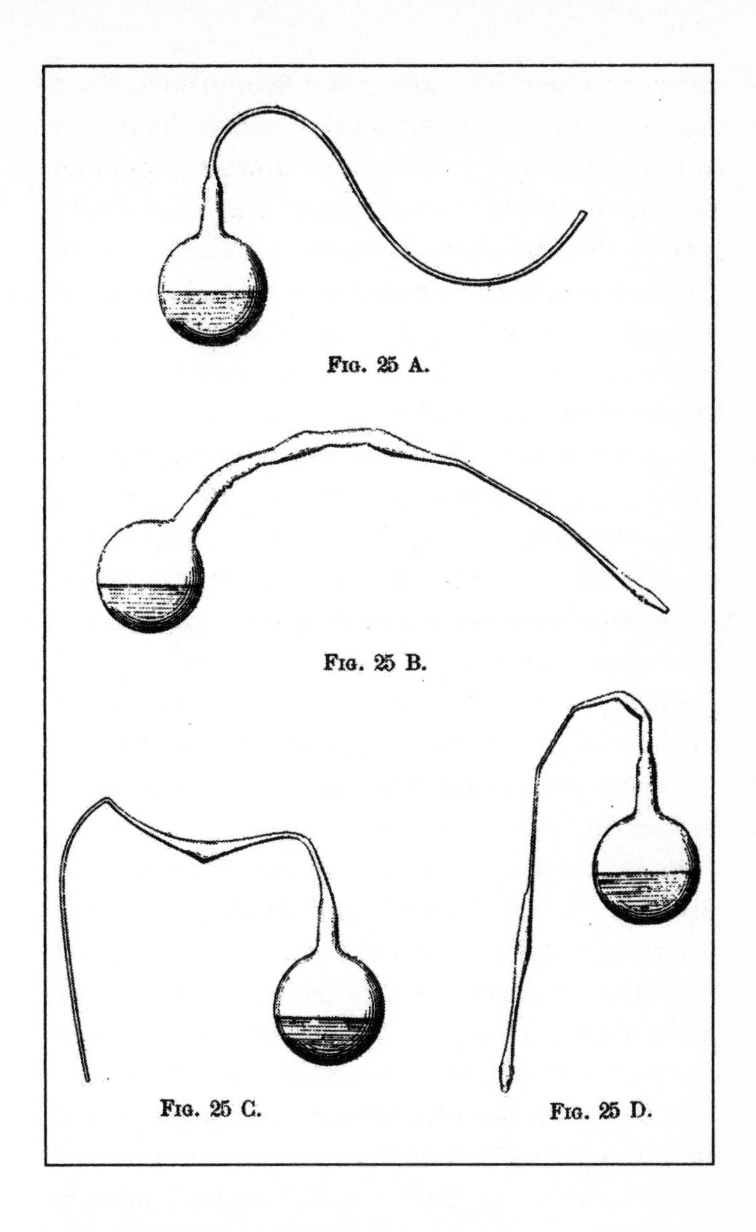

Illustration 4: Pasteur's swan-necked flasks. From *Oeuvres de Pasteur*, Vol. 2, Paris, 1922.

were contorted in a variety of different ways. These contortions impeded, to varying degrees, the flow of airborne matter into his flasks but did not inhibit the passage of oxygen. As Pasteur expected, the easier it was for airborne matter to reach the sugar-water, the greater the degree of fermentation that occurred, but where the germ-bearing dust particles became trapped in the convolutions of flasks with twisted necks, no fermentation took place.

When Pouchet, like Needham, then claimed that heating organic solutions destroys their capacity to generate new life, Pasteur found a way of extracting sterile fluids from cattle that didn't need to be heated before they were hermetically sealed, incubated and later examined for evidence of microbes. Again, neither germs nor putrefaction were visible several weeks later. This was experimental science at its best and once again Pasteur revealed himself to be a master practitioner.

But what Pasteur didn't admit at the Sorbonne was that many of his own experiments had actually seemed to confirm Pouchet's findings. We know now that this is because even Pasteur could not always prevent contamination. But during the 1860s there was no objective way for him to decide whether the flaw lay in his prior belief that spontaneous generation is nonsense or with his experimental apparatus. Indeed, recent investigations of Pasteur's notebooks show that from the very start he was utterly committed to disproving spontaneous generation. At least in part, this is because his conservative religious principles had convinced him that only the Almighty could create new life. Certain of this notion, he jotted 'successful' whenever microbes failed to appear in his flasks, and 'unsuccessful' when they did appear. Spontaneous generation not only seemed like a bad scientific idea to Louis Pasteur, it conflicted with his religious sensibilities as well.

While Pasteur's experiments were persuasive and beautifully ingenious, there was also one critical test that he failed to perform. Despite being almost 80, in 1864 Felix Pouchet had climbed high into the Pyrenees and exposed a flask containing a sterile hay-infusion to the open air. Even though Pasteur's own experiments had shown that mountain air is virtually germ-free, Pouchet's flasks later showed emphatic evidence of microbial life and he concluded once more that oxygen and organic matter are all that are needed for life to come into being. In response, Pasteur simply impugned his rival's methods.

Yet, like Pouchet, Pasteur was working on the false assumption that a few minutes of boiling will kill all microbes. Had he attempted to replicate Pouchet's experiment, he may well have found himself in something of a quandary. It was later shown that hay often contains bacteria that can be killed only by a process of repeated boiling and cooling. Pouchet's hay is very likely to have been contaminated with such bacteria. Difficult though it is to admit, it now seems indisputable that Pasteur's failure on this occasion to play fair actually served the advancement of science. By departing briefly from the narrow path of the scientific method, and refusing to repeat Pouchet's Pyrenean experiment, he – and the wider world – were saved a tremendous amount of confusion.

As it was, the French scientific community had had enough of Pouchet by 1864. And this too had as much to do with religion as with science. Since the late 1700s, spontaneous generation had been strongly linked to the heretical claim that mankind, and all other complex organisms, had evolved from spontaneously generated microbial organisms called 'monads'. In the deeply conservative religious and political climate of 1860s France, it didn't take much to convince scientists that Pasteur was in the right and Pouchet a dangerous heretic. Once an officially appointed scientific commission of conservative French scientists had voted against Pouchet for a second time in 1865, he sloped off to Rouen an embittered man. Pouchet may have suffered from something of a persecution complex; but there's no disputing the fact that he quit Paris knowing he had lost a less than fair fight.

Tyndall and Cohn

Following Pouchet's departure, the French mostly abandoned spontaneous generationism. In Britain, too, its supporters' lives were now being made more and more difficult.

The Irish-born physicist John Tyndall had already amazed audiences by illuminating the air with concentrated beams of light and revealing the large quantity of 'floating matter' it carries. Tyndall then made a vital conceptual link. Asserting that there are strong similarities between human disease and the breaking down of organic substances during the biological process of fermentation, he pointed into the beam of light and declared that every one of the hundreds of thousands of dust particles his audience could see carried potentially deadly germs. Tyndall had made a leap in the dark but had landed on solid ground.

In 1876, in an attempt to silence the remaining spontaneous generationists, he designed a small, closed chamber with two unusual features. First, test tubes of organic matter were inserted through the bottom. Second, its inner faces were smeared with a sticky glycerine mixture. Having left the chamber undisturbed for a few days, Tyndall shone a beam of light into it and showed that all the airborne particles had either settled to the bottom or been captured by the chamber's sticky surface. He then boiled the contents of the test tubes and left them to cool with their necks open to the inside of the chamber. As he had anticipated, the organic matter was still sterile several days later. Oxygen and organic matter, his experiment indicated, are not enough to produce microbes; for this the airborne particles are essential, and they had no longer been available to seed the test tubes.

But Tyndall's biggest contribution to the debate came the following year. Pouchet and the English doctor H. Charlton Bastian had both observed putrefaction in materials even though they had been carefully boiled and kept away from atmospheric air. This had convinced them that spontaneous generation really happened. Tyndall's master-stroke was to show that some germs are far less easily destroyed than everyone had previously thought. Some have a capacity to resist heat and can be killed only by alternate boiling and cooling. In other words, Tyndall showed that Pouchet and Bastian had discovered the existence of heat-resistant bacteria but had failed to realise it. These findings were soon after confirmed, and greatly refined, by the German bacteriologist Ferdinand Cohn. In 1876, Cohn discovered the bacteria *Bacillus subtilis*, which Pouchet's hay infusion had almost certainly contained. This incredibly tenacious germ survives extremes of temperature, he demonstrated, by forming thick-skinned spores. For those prepared to listen, Cohn's capstone discovery now destroyed the last vestiges of Pouchet's case.

· CHAPTER 9 ·

THE ENGLISH DISCIPLE

Joseph Lister had the advantage of beginning his surgical career more than three decades later than Ignaz Semmelweis. This meant that, while he did meet with substantial opposition, he lived long enough to see the ideas of contagion and germ theory triumph and his services to medicine rewarded with a baronetcy. But none of this could have been foreseen when, in 1854, this son of a well-to-do English wine merchant became an assistant surgeon at Glasgow Royal Infirmary.

Having already established a considerable reputation in London as a physiologist, Lister had decided to switch to surgery and he moved to Scotland to obtain the best available training. Although initially sceptical of germ theory, in Glasgow he was won over by Pasteur's 'beautiful researches' on the role of microbes in causing fermentation and decay. Like Tyndall, he then made a connection between Pasteur's test tubes and the ghastly wound infections that killed so many of his patients. If germs can cause putrefaction in meat, Lister reasoned, then they might also be responsible for the grim rates of post-operative death in his wards.

This radical intuition told Lister that the key to reducing post-operative infection was to keep airborne germs away from his patients' wounds. On 12 August 1865, he put his theory into practice. A young boy called James Greenlees had received a compound tibia fracture when run over by a cart; the high rates of infection at the time meant that his prognosis was very poor indeed. So Lister soaked the boy's wounds in an antiseptic called carbolic acid and then covered them with dressings impregnated with more carbolic acid. 'Hospital gangrene' didn't put in an appearance and, six weeks later, Greenlees left the hospital, his leg fully healed.

A few years later, Lister added a carbolic acid spray to his surgical repertoire. During operations, one assistant constantly pumped carbolic acid into the air, producing a fine antibacterial mist. Then, with the application of carbolic acid dressings, the patient's wounds were kept drenched in antiseptics for the entire period of convalescence. With these innovations, the death rate in Lister's wards began to fall. As it did so, he extolled the power of antiseptics in dozens of medical books and articles. Within a few years, he was able to boast of a 30 per cent reduction in the rate of post-operative mortality. And, as news of his successes spread, Lister acquired a burgeoning fan club.

But this was no overnight revolution. From the

start, Lister met with stiff opposition, and this was by no means always irrational. It is often overlooked that Lister was far less original than he and his disciples asserted. He was only one of dozens of surgeons during the 1860s dedicated to making surgery safe. Having already imbibed the ideas of Holmes, Semmelweis, Callender and Nightingale, it was common practice for many hospitals to separate surgical patients, ensure good ventilation and use antiseptics (including carbolic acid) in cleaning wards – reforms all made without the aid of either Pasteur or germ theory.

More strikingly, surgeons like Callender, who refused to employ carbolic acid dressings and sprays, but who kept their wards scrupulously clean, boasted considerably higher survival rates than Lister. This is largely because, until the 1880s, Lister believed germs to be an airborne threat alone. As a result, unlike many other surgeons, he made little effort to keep his wards disinfected and free from surgical detritus. One visitor made the following diary note in 1871: 'Although great care is evidently taken to carry out the antiseptic treatment so far as dressings are concerned – there is a great want of general cleanliness in the wards – the bed clothes & patients linen are needlessly stained with blood and discharge.' Another guest wrote that Lister habitually 'wore an old blue frock-coat for operations, which he had previously worn in the dissecting room', and which was 'stiff and glazed with blood'.

Despite these failings, however, Lister did play an

essential role in gaining acceptance for the germ theory. Not only did his proselytising ensure that all surgeons were at least familiar with Pasteur's work, but his skilled public relations led many Continental surgeons to take up his methods and ideas with a rigour that ensured their success.

Among his earliest converts was the Swiss obstetrician Johann Jacob Bischoff. Having visited Lister's wards in 1868, Bischoff returned to Basle and within a decade had reduced mortality rates from childbed fever by 80 per cent. No other hospital up till then had achieved such a startling reduction in mortality. In 1870, brimming with enthusiasm after meeting Lister, Denmark's A. Stadfelt was just as effective at bringing down death rates in his Copenhagen lying-in hospital. Before taking part in a delivery, midwives were placed in a room in which they breathed fresh air through a tube while their bodies and clothes were exposed to sulphuric acid fumes for fifteen minutes. Then, before and after each delivery, hands, instruments and catheters were carefully disinfected. Mortality from childbed fever fell by more than two-thirds. This was an incredible achievement.

During the Franco-Prussian war, deaths from gangrene in both sides' military hospitals were distressingly high. But rarely were they so extreme as on the wards of the Munich clinic run by Johann Ritter von Nussbaum, where the mortality rate had reached 80 per cent by the early 1870s. In desperation, Nussbaum sent one of his assistants to Glasgow, and he returned with news of Lister's method. 'Behold now my wards,' Nussbaum enthused a few years later, 'which so recently were ravaged by death. I can only say that I and my assistants and nurses are overwhelmed with joy.'

By the early 1880s, many other German surgeons had adopted antisepsis with similar rigour. Nussbaum famously declared that a doctor who examined a wound before disinfecting his fingers ought to be charged with criminal negligence. And in many German hospitals, surgeons were now forbidden from attending post-mortems or infectious patients for 24 hours prior to performing surgery. In addition, outside clothing was either removed or covered with clean aprons before operations began; hands and arms were washed with carbolic acid; and a nailbrush was used to scrub the hands and arms as well as the nails.

Other preventative measures were also introduced into Germany. Prior to surgery, both operating theatres and patients were liberally soaked in antiseptic. Operating tables were constructed of a single slab of glass so they could easily be washed. And surgical instruments were manufactured from solid pieces of metal so that germs were unable to collect in the gaps between the handle and the blade. Lastly, basins of carbolic solution were at all times kept on operating tables so that the surgeons' hands and instruments could be continually re-washed.

The success of these precautions provided undeniable evidence for the contagiousness of hospital infections. And, paying homage to Lister as the originator of antiseptic surgery, by the late 1870s doctors and surgeons also began to take serious note of his claim that living germs cause infection.

The gospel was spreading.

Part IV

Worms, Chickens and Sheep

· CHAPTER 10 ·

THE PLIGHT OF THE SILKWORM

From the privileged vantage point of the present, it seems obvious that having defeated Felix Pouchet, Louis Pasteur should have turned his attention to contagious diseases. We have already seen how Lister and Tyndall inferred from his experiments that if germs cause rotting and decay, then they might cause illness too. In an 1859 paper on the subject of fermentation, Pasteur had remarked that 'everything indicates that contagious diseases owe their existence to similar causes'. But he had taken this idea no further.

Instead, despite the fact that he now spent five years investigating a devastating silkworm disease caused by microbes, Pasteur struggled to make the connection between germs and illness. Yet, these five frustrating years were crucial in his growth as a scientist. For having finally arrived at the correct answer, he would never again doubt the role of germs in causing infectious disease. The result would be among the most remarkable series of discoveries ever recorded in the history of science.

By the 1860s, the silkworm industry of southern France was more than six centuries old. It had been hard hit during the Revolution because the revolutionaries had not shared the Bourbon monarchy's taste for ostentatious fabrics. But the subsequent imperial and monarchical regimes had a strong predilection for fine silks and, with Napoleon III as the head of state, the industry seemed set for a period of great prosperity. Then a disaster of a seemingly less reversible nature struck. A disease called *pébrine* began to cause millions of French silkworms to shrivel up and drop off the mulberry twigs on which they had woven their silky cocoons for hundreds of years.

With almost palpable horror, in 1865 the region's senator Jean-Baptiste Dumas wrote to Pasteur imploring him to help his 'poor land'. 'Its state of misery', Dumas lamented, 'is beyond anything you can imagine.' Flattered and moved in equal measure, Pasteur agreed to assist. Although having previously 'never touched a silkworm', he devoted every summer for the next five years to studying this devastating disease: the humble silkworm found a powerful ally.

But having journeyed to the South of France with his loyal assistants, Désiré Gernez and Emile Duclaux, Pasteur found that he would have to share the limelight. One of the now largely forgotten architects of modern germ theory was already hard at work. Antoine Béchamp, a noted scientist in his own time, had recently discovered the presence of what looked like microbes on the exterior of diseased silkworm eggs. His research had been inspired by the Italian estate-owner Agostino Bassi who, during the 1830s, had shown that another fatal disease of the silkworm, muscardine, is caused by a parasitic fungus. That was the first time a tiny organism had been shown to cause a deadly illness. Three decades later, Béchamp asserted that *pébrine* too is a contagious disease caused by tiny parasites that burrow inside the silkworm eggs and doom generation after generation to premature death.

Pasteur soon found out about Béchamp's research and he was moved to disparage his rival's explanation as a 'bold-faced lie'. Even when Béchamp showed that the parasite-like corpuscles he had found on silkworm eggs caused cane juice to ferment, Pasteur continued to lambast his rival with a savagery unusual even for him. *Pébrine,* he declared, is not contagious but hereditary and the corpuscles are merely decomposing silkworm cells. Seemingly content to play Liebig to Béchamp's Pasteur, he argued with angry obstinacy that the corpuscles are merely an effect, and not the cause, of the disease; thus even the man who had first speculated on the parallels between fermentation and disease failed to see the significance of Béchamp's results.

Luckily, Pasteur's assistants were more sympathetic to Béchamp's ideas. Gernez, in particular, had taken great strides in proving him correct. In one experiment, mulberry leaves free from the excreta of silkworms with *pébrine* were fed to healthy worms. The result was unambiguous: they all produced cocoons without any sign of the disease. Gernez then ground up leaves from a tree contaminated with worms that had died of *pébrine*. This gave him a paste that was painted onto a batch of mulberry leaves and fed to a second set of healthy worms. They, in contrast, wove very few cocoons and those that they did were riddled with corpuscles. In another series of experiments, Gernez fed normal silkworms a rich broth containing the mysterious corpuscles. Sure enough, the few worms that clung to life long enough to weave cocoons left behind a legacy of hundreds of diseased eggs.

But still Pasteur was unconvinced. Nor was he being entirely unreasonable; for some of his own silkworms had become very unwell before the corpuscles had put in their appearance. How, therefore, could the corpuscles have caused the disease?

Then, in early 1869, Pasteur suddenly saw his error. Without realising it, he had been studying two entirely different illnesses. The worms that had fallen ill before the corpuscles appeared had actually first been struck down with *flacherie*, a very different infection. Only later, when the corpuscles appeared, had the exhausted worms fallen prey to *pébrine* as well. The appearance of the corpuscles did after all coincide with the onset of *pébrine* and, at last, Pasteur was ready to accept that Gernez, Duclaux and Béchamp had been right all along.

But even before this flash of insight, Pasteur had helped introduce the practices necessary to banish *pébrine* from the orchards of southern France. Believing the disease to be inherited, in 1866 he had explained to farmers how they must carefully select the eggs from which to grow each new generation of silkworms. Once the females had mated and laid their eggs, they were to be dissected and microscopically examined for traces of corpuscles. If corpuscles were found, farmers were instructed to burn the eggs immediately.

Naturally, this eugenic method was equally appropriate for tackling microbial disease: the worms with *pébrine* corpuscles were destroyed, and the infection with them. Thanks to this simple innovation, and inspired by the ability of Pasteur's young daughter to operate a microscope, many of the region's farmers began once more to have bumper harvests of healthy silkworm eggs. As one might expect, however, some fiercely independent local farmers were unprepared to see ancestral practices overturned and they carried on much as before. Perhaps a few even went to the grave with their fists raised in defiance against the Parisian interloper. But most fell into line over the next few months as, with a skill that seemed to border on the clairvoyant, each of Pasteur's bold prophecies about whether individual batches of eggs would produce healthy or diseased silkworms were proven unequivocally correct. In one case, in March 1869, he sent four packets of eggs to the Silk Commission of Lyon with a set of exact predictions as to what their state of health would be upon hatching. In every case he was spot on.

Pasteur finally left Alais in the South of France partially paralysed by a stroke, but his years studying silkworms had been very worthwhile. Between 1865 and 1870, he had received an object lesson in the role of microbes in causing disease. Although Pasteur would never have the grace to admit it, Béchamp had given him a powerful shove in the right direction. Infectious disease was now to be his sole interest. From it his greatest achievements were about to come.

· CHAPTER 11 ·

ANTHRAX

As Pasteur toiled among the orchards of Alais, closer to Paris another Frenchman had begun studying a very different disease, and one that would play a vital part in unlocking the secret of germ theory. Casimir-Joseph Davaine was everything that Pasteur was not: modest, unprepossessing and quickly forgotten by posterity. But he shared with his countryman the tenacity required to advance the germ theory a crucial few steps further.

Between 1863 and 1870, Davaine single-mindedly struggled to prove that anthrax is a microbial disease. Responsible for thousands of livestock deaths every year, anthrax had for long been a puzzle. As it would appear spontaneously and destroy entire flocks that had had no apparent contact with other anthracic sheep, standard models of contagion or miasma just didn't apply. As such, most veterinarians had long since decided that some combination of soil and humidity must be to blame.

Davaine's vital contribution built on the observations of earlier French and German scientists who, under the microscope, had noticed the presence of 'stick-shaped corpuscles' in the blood of anthracic sheep. But it was only after having read Pasteur's work on fermentation, in 1863, that Davaine became convinced that these corpuscles were the sole cause of the disease. Seeking to confirm his hunch, he transferred blood from an infected sheep to healthy rats and rabbits. Three days later he was elated to find that all the animals were dead of anthrax. Surely, he reasoned, he had furnished an absolute proof of the germ theory of anthrax. But Davaine was in for a disappointment. It rapidly became clear that his first experiment had been a lot less compelling than he thought.

This is because, as his critics were quick to point out, Davaine had failed to isolate the stick-shaped corpuscles from all the other fluids and solids extracted with the sheep's blood. It was therefore just as likely that something else in their blood had caused the animals to succumb to anthrax. His experiment had proven only that the disease agent had something to do with the vascular system.

In response, Davaine undertook a series of exceptionally elegant experiments. In one, he varied the number of microbes in the blood that he injected into laboratory animals and found that the higher the concentration of stick-shaped corpuscles, the faster the animals died. In another, he diluted some anthracic blood in distilled water and let it stand until the bacteria had settled to the bottom of the flask. Using a syringe he then injected a series of guinea pigs with either clear water from the top of the flask or the muddier, microbe-rich water taken from the bottom. Sure enough, those guinea pigs injected with the water from the top of the flask enjoyed a brief reprieve.

This was persuasive stuff and by 1870 the idea that anthrax is caused by microbes was making real headway. Yet for all his experimental flair, Davaine had by no means proven that microbes are the cause of anthrax. Completely isolating the stick-shaped corpuscles had been beyond his technical ability. More significantly, he still couldn't explain the most curious feature of the disease's epidemiology: the way in which flocks of perfectly healthy animals could be wiped out almost overnight by anthrax, despite their never having coming into contact with other infected animals.

In an obscure town in Eastern Germany, a general practitioner named Robert Koch, with a home-made laboratory in his back room and a penchant for microscopy, was among the many who felt that Davaine's work remained unfinished.

Enter Robert Koch

Born in the town of Clausthal, in Germany's Harz mountains, from an early age Robert Koch displayed the qualities that would later win him fame: extraordinary industry, patience and perseverance. He was also fortunate to have briefly studied medicine under

Jacob Henle at the University of Göttingen. In a paper of 1840, Henle had reviewed the fragmentary evidence for the role of germs in causing disease and concluded that their importance was potentially very great indeed. Not least because Henle fell far short of proving his case, few had taken much notice. But Koch must have been aware of his mentor's paper and he almost certainly left Göttingen in 1866 more prepared than most doctors to take germ theory seriously.

In the following years, Koch and his new wife moved further and further eastwards towards the Polish border in search of a lucrative private practice. By the early 1870s, after a brief stint as a surgeon during the Franco-Prussian War, Koch had settled in Wollstein, a town in which anthrax happened to be rife. In 1873, intrigued by Davaine's results, he obtained some anthracic blood and proceeded to maintain an infection through twenty successive generations of mice. Reassured that Davaine was probably on the right track, he now applied his considerable mental powers to the question of how outbreaks of anthrax could come and go in such an inexplicable manner. This was the real sticking point for the germ theory of anthrax and, between lancing boils and pulling teeth, Koch worried at it relentlessly.

He reasoned that if Davaine's rod-shaped corpuscles were the true cause of anthrax then they must be able to survive for long periods of time outside of the animal body. He could imagine only one way of their

Illustration 5: A portrait of Robert Koch. Anonymous lithograph. Source: The Wellcome Library, London.

doing so: they would have to turn into spores that could lie dormant until ingested by a passing animal. In order to test this hypothesis, Koch took some more anthracic blood and cultivated the microbes in the aqueous humours of ox-eyes spread upon microscope slides. With extraordinary patience, he then began to observe the mysterious unfolding of life beneath his cover slip.

By the time he had finished, Koch had made a gratifying discovery. The microbes had a life cycle fully compatible with their being the cause of anthrax. Every so often, they produced spores that looked like tiny strings of pearls and were clearly able to withstand extremes of temperature and exposure. But upon the addition of more aqueous humour, they developed into the rod-shaped bacilli that were always present in anthracic mice and sheep. Koch surmised that anthrax spores typically lie in grass until consumed by grazing sheep. Upon finding themselves in nutrient-rich environments they then germinate, begin to metabolise and breed, finally destroying their host and leaving behind millions of tiny spores ready for their next victim.

Taking the obvious next step, Koch injected some of these spores into a batch of mice. They quickly developed anthrax and died. But Koch was not yet satisfied, for he still didn't know for sure whether the anthracic bacilli were a distinctive type of bacteria capable only of causing anthrax or whether one germ was much like any another. Koch had to demonstrate that there are different types of bacteria and that some of these cause illness even if others don't. He achieved this by searching out another spore-producing microbe, which when it was injected into mice produced consistent results quite different to those resulting from the anthrax spore. Specific microbial agents, he concluded, cause specific bodily infections. In his own mind, at least, the germ theory of disease was finally established. From Davaine's initial insight, an increasingly robust scientific theory was emerging.

An Unknown Physician

In the spring of 1876, Koch wrote with a curious blend of fawning and self-assurance to the expert bacteriologist, Ferdinand Cohn, at the University of Breslau.

> *Esteemed Professor, Stimulated by your bacterial investigations ... I have been investigating the aetiology of anthrax for a considerable time. After many failures, I have finally succeeded in completely elucidating the developmental cycle of* Bacillus anthracis. *I believe I have amply confirmed my results. Nevertheless, esteemed Professor, I would be most grateful if you, as the leading authority on bacteria, would give me your criticism of my work.*

Cohn had received many such letters from misguided 'dilettantes' in the past and expected little from this 'unknown physician from a rural Polish town'.

Luckily, however, he agreed to Koch's proposed visit. And, within days, the humble doctor arrived in Breslau laden with crates of equipment and animals.

Years later, Cohn described how 'during the very first hour I recognised that he was the unsurpassed master of scientific research'. Because so little was expected from the obscure provincial, few were present on the first day to affirm Cohn's judgement. But news spread fast. By the second day of Koch's visit, more and more leading doctors were turning up for his demonstrations. At the end of day two, the Director of the Pathological Institute told a colleague, 'Drop everything and see Koch.' On day three, the audience comprised most of the University's foremost scientists. At the close of this final session, one of them presciently remarked: 'Koch will continue to surprise us and shame us with further discoveries.'

Returning home in triumph, Koch wrote up and published his results, bringing himself to the attention of an international audience. Even so, he was obliged to go on seeing private patients for another five years until his celebrity secured him the full-time research position he so richly deserved. In the meantime, his critics were utterly vanquished.

Their line of attack was to point out that even Koch had been unable to separate entirely the bacteria from the blood of the animals from which it had been obtained. It therefore remained possible that it was something in the blood, and not a specific bacterium, that caused anthrax. Within months, this claim had been brilliantly refuted by a man soon to become Koch's arch-rival: Louis Pasteur.

An Encore at Pouilly-le-Fort

In 1876, Pasteur intervened to flush the opponents of the germ theory of anthrax from their last redoubt. He first managed to grow the anthrax bacilli for several generations in a sterile culture of urine before injecting it into guinea pigs. This time no blood was transferred and, confounding the critics, still the animals developed anthrax and died. Then he travelled down to the beautiful Eure-et-Loire and studied how animals contract anthrax in natural conditions.

In one experiment, several healthy sheep were fenced into an enclosure above a deep grave containing the corpse of a sheep that had died of anthrax ten months earlier. Before long, one of these sheep contracted and perished from the disease. Exchanging microscope for trowel, Pasteur collected numerous earthworm casts from inside the enclosure, which he then examined in the laboratory. Several of them, he was delighted to find, were riddled with anthrax spores.

The best, however, was yet to come.

On Thursday 2 June 1881, Pasteur launched himself into the scientific stratosphere by undertaking one of the most daring public experiments ever performed. He had recently announced his discovery of a vaccine against anthrax. Shortly after, aware of

Pasteur's weakness for the grand public spectacle, a veterinarian named Hippolyte Rossignol had invited him to test its efficacy in an open forum. As the new vaccine comprised attenuated (i.e. weakened) versions of the *Bacillus anthracis*, all participants realised that this was to be as much a test of the germ theory as of the vaccine itself.

To those who knew anything about Rossignol, the motivation behind his challenge was obvious. On several occasions, this self-assured veterinarian had publicly repudiated the germ theory of anthrax, sarcastically referring to Pasteur as its 'high priest' and 'prophet'. In hosting a public trial at his farm in the town of Pouilly-le-Fort, his hope was that the inadequacies of germ theory would be laid bare for all to see. Pasteur, in short, was to be provided with every opportunity for hanging himself on his own rope.

The experiment was conducted according to a strict scientific protocol. On 5 May, 50 healthy sheep were set aside. Twenty-five were then injected with Pasteur's anthrax vaccine and the remainder marked as controls. On 17 May, the first set of sheep received another 'protective injection'. Finally, on 31 May, all 50 were injected with a virulent strain of the supposed anthrax microbe. Come 2 June, Pasteur boldly prophesied, the 25 vaccinated sheep will be the sole survivors.

On the day in question, representatives of the world's media converged on Rossignol's modest farm, eager to get news of the trial's results to impatient editors. Neither germ theory nor the use of vaccines was sufficiently well accepted for the result to be seen as a foregone conclusion. Indeed, just a few days prior to this day of reckoning, Pasteur had received a telegram suggesting that some of the vaccinated sheep were looking distinctly unwell. The Frenchman was playing for the highest possible stakes. And, for a moment, he sensed public humiliation, lost his nerve, accused his team of incompetence, and barked that someone would have to go to Pouilly-le-Fort in his place.

But on the big day, the sun shone on Louis Pasteur. All but one of the vaccinated sheep had survived, the other perishing because of complications with the foetus she was carrying, and all 25 of the controls lay dead or dying with unmistakable cases of anthrax. The trial had been an unmitigated triumph. When Pasteur stepped off the train from Paris and headed for Rossignol's farm, he was greeted by jubilant crowds. The role of specific germs in causing anthrax had been decisively proven and Pasteur had a vaccine that would save farmers millions of francs and fill the coffers of his own laboratory.

Yet that Pasteur monopolised the accolades of June 1881 was not entirely just. For if we look beyond the standard textbook account, we find a more complex story. To suggest that this was Pasteur's victory alone is to misrepresent the events that led up to the Pouilly-le-Fort trial. In fact, the anthrax vaccine emerged out of almost a decade of intense collaboration in which

Pasteur was only one, and perhaps not even the most important, participant. To really understand the events of 2 June 1881, we need to go back to the late 1870s and to the discovery of the very first laboratory vaccine.

The First Laboratory Vaccine

The principle of immunity had been established in ancient times. 'The sick and dying were tended by the pitying care of those who had recovered', wrote the historian Thucydides of the plague epidemic that devastated the Greeks during the Peloponnesian Wars, 'because they knew the course of the disease and were themselves free from apprehensions. For no one was ever attacked a second time.' As we've seen, by the eleventh century, smallpox vaccinations had been introduced in the Far East. Encouraged by their effectiveness, the nineteenth-century French pioneer germ theorist, Dr Joseph Auzias-Turenne, would spend years hunting for a way to produce a vaccine effective against syphilis.

Despite years of effort, Auzias-Turenne failed. But in paper after paper he stressed one of the basic ideas underlying the search for vaccines. 'One virus can take on different forms and manners', he explained, 'The viruses regenerate or strengthen; they degenerate or weaken.' Pasteur surely knew of Auzias-Turenne's work. He was also, of course, aware of Jenner's discovery that milder cowpox confers protection against smallpox. That some individuals have, or can acquire, agents that fight disease was also underscored by dozens of observations in Pasteur's own laboratory demonstrating that animals have differing susceptibilities to the same microbes. Still, by the late 1870s smallpox vaccination remained an isolated medical victory. Then, in the spring of 1879, having identified the microbe responsible for chicken cholera, Pasteur began to seek a way of reducing its virulence.

According to legend, Pasteur forgot about a culture of chicken cholera microbes and then, quite by chance, several months later injected it into a batch of chickens. Surprised to find the chickens all unscathed after a few days, he next gave them injections of full-blown cholera bacteria. When they survived this challenge too, Pasteur realised that simply exposing the bacteria to air is enough to cause attenuation.

In reality, as his laboratory notebooks have recently revealed, this decisive experiment never took place. It actually took Pasteur and his team several months to register that if they left broths containing chicken cholera undisturbed for a few weeks, the virulence of the bacteria declined. There was no sudden moment of illumination. And not until a year after the legendary experiment is supposed to have taken place did Pasteur recognise that the bacteria had been weakened by exposure to oxygen.

Furthermore, the breakthrough that eventually led to the first laboratory vaccine was due not to Pasteur himself, but to his principal assistant, Emile Roux.

Roux was much less of a yes-man than most of Pasteur's subordinates. Notoriously fiery tempered, he had been kicked out of the army's medical school for insulting a general who expressed doubts as to the value of his research. Soon after, driven by poverty to become a schoolteacher, he almost asphyxiated a disobedient student. Eventually he landed a humble job at Paris' Hôtel-Dieu. And there his luck changed. He found an old teacher who was now working with Pasteur and who happily invited him to join the team.

Yet, despite working for the eminent Pasteur, Roux relinquished little of his independent spirit. And it was while conducting his own inquiries that he had the idea of passing a stream of oxygen across the chicken cholera broth. Why it worked, he didn't understand. But, by the middle of 1880, the first laboratory vaccine had been discovered. From then on, Pasteur's efforts were almost entirely directed at vaccine research. And it was at this stage that he turned his attention to anthrax.

Here, though, Pasteur was in for a rude shock. In July 1880, the French veterinarian Jean-Joseph Henri Touissant announced that he had developed an effective anthrax vaccine by heat-treating a liquid solution containing live anthrax microbes. Pasteur was alarmed to hear that he had been beaten to the line. But he was reassured when it transpired that Touissant's vaccine was less than reliable. Seeing that the race was not yet lost, Pasteur redoubled his own efforts and demanded that his colleagues cancel their summer vacations. At last, in February 1881, he finally declared that he and his team had succeeded where Touissant had at least partially failed.

Because anthrax bacteria form highly resistant spores, developing a vaccine was a much more difficult proposition than it had been with chicken cholera. In this case, simple air exposure wasn't enough. Casting about for an alternative, Pasteur was presumably inspired by Touissant's method of heat treatment to search for a temperature at which spores did not form and therefore allowed the bacteria to be weakened. With exceptional technical skill, and an exactitude absent from Touissant's research, he eventually found that at 42°C anthrax bacteria begin to attenuate.

It was at this juncture that Rossignol threw down the gauntlet. Pasteur agreed to his terms with apparent alacrity. Privately, however, he must have felt a strong sense of foreboding. For, unbeknown to the public, his vaccine was far from ready. Despite weeks of careful research, it was still little more reliable than Touissant's, about which he had been utterly dismissive. Only three weeks before the trial, a small-scale test suggested that public humiliation was almost certainly on the cards.

Fortunately for Pasteur, in early 1881 his team got wind of an improvement that Touissant had made to his method of attenuating bacteria. A decade earlier, Davaine had shown that certain antiseptics can destroy the anthrax bacillus. In August 1880, Touissant built on this insight and soaked anthrax bacteria in carbolic acid in an attempt to weaken them. His results were very promising and, after a short delay, he rather generously divulged details of his new method to Roux, who passed the information on to his boss. Now, while Pasteur continued to work on his heat-attenuated vaccines, one of his assistants, Charles Chamberland, experimented with scores of different chemicals to find which was most adept at weakening the anthrax bacilli. He eventually found that washing anthrax spores in potassium-bichromate gave the best results.

Contrary to the standard account, at Pouilly-le-Fort it was Chamberland's vaccine, not Pasteur's, that was used to such acclaim. Only in the months following the trial did Pasteur have the chance to refine the use of heat. Then it gradually emerged as the premier method for producing anthrax vaccine. Both methods, however, owed much to Touissant. He had had neither the training nor the resources to make the best use of his discoveries. But brilliantly exploited by Pasteur's team, his work had been transformed into a major scientific success.

This account of the development of the first anthrax vaccines, based on the letters and notebooks of those involved, underscores the profound importance in science not just of the competitive instinct, but also of the rapid exchange of ideas, inspiration and technical know-how. Pasteur was by no means always a front-runner, sprinting ahead of the pack and leaving colleagues and competitors in the dust. Without Touissant's and Roux's breakthroughs, the first reliable anthrax vaccines might well have been developed elsewhere.

This observation is not intended to dislodge Pasteur from his pedestal. Most episodes in the history of science testify to the far greater role played by collaboration and cross-fertilisation than by individual initiative, in achieving scientific progress. In the majority of cases, the popular notion of the lone genius who heroically overcomes envy, ignorance and anti-rationalist church dogma in pursuit of the truth is little more than a romantic cliché. Good science, in short, is a collective enterprise. Pasteur, nonetheless, was a man of quite exceptional talent.

Part V

Koch's Postulates

· Chapter 12 ·

1881: Potatoes and Postulates

During the late 1870s, before the Pouilly-le-Fort trial, Koch had turned to the study of infective diseases such as septicaemia and gangrene in rabbits and mice. He claimed to have identified six specific bacteria that always caused identical deadly infections in their hosts. What made this work especially significant was that all six closely resembled human diseases. In retrospect, this research marks a major change of tempo in the germ revolution. Hitherto progress had taken the form of individual thrusts, as single diseases were brought within the ambit of germ theory. Now Koch was not only switching to a broad front, he was advancing on it at a prodigious rate. Soon no sensible doctor would be able to disavow entirely the role of germs in causing disease.

Yet few anticipated the scale of the transformation that Koch would inaugurate. This is unsurprising. As the microbes responsible for the biggest and nastiest epidemic diseases had yet to be found, few were willing to extrapolate from a handful of animal diseases to every conceivable form of infectious illness. To most doctors, miasmatist explanations for human infectious illness were far more compelling. The germ theory itself seemed to have a very limited application.

The 'Blind Zeal' of Louis Pasteur

Koch's strategy in attacking the serried ranks of doubters was masterly. In 1881, he carefully noted all the objections to germ theory. Then he publicly set out what was required in order to prove once and for all the role of specific germs in causing specific diseases. The criteria he set became famous as Koch's four postulates. First, the bacterium must be present in every case of the disease. Second, the bacterium must be isolated from the diseased host and grown in pure culture. Third, the specific disease must be reproduced when a pure culture of the bacterium is inoculated into a healthy, susceptible host. Lastly, the bacterium must be recoverable from experimentally infected hosts. To bacteriologists, these postulates soon acquired an authority on a par with the rules Moses brought down from Mount Sinai; fortunately, as techniques improved, they would become an awful lot easier to observe.

At the beginning even Koch acknowledged that isolating specific microbes from the bodies of the victims of disease was a very tall order indeed. And, not least because he was now embroiled in a bitter personal rivalry with Pasteur, he was happy to admit that many of the microbial cultures previously used had been far too contaminated to validate the single-germ single-disease hypothesis. 'On the whole,' Koch icily observed, 'it is truly depressing to attempt pure cultivations', a remark he considered 'especially applicable to the researches (carried on with really remarkable, if blind, zeal) now issued in quantities from Pasteur's laboratory, and which describe incredible facts with regard to pure cultivations'.

This was outrageous effrontery from a man who had been a schoolboy when Pasteur had put Pouchet to flight. When Koch soon after added that the Frenchman had contributed 'nothing new to science', a relationship already profoundly damaged by the Franco-Prussian war went off the negative end of the scale. It would never recover. Nevertheless, from a technical perspective, Koch did have a point. Even though both he and Pasteur claimed to have identified distinct microbes in the past, for good reasons many scientists remained entirely unconvinced.

A key problem was the extreme difficulty of telling apart different species of bacteria. In 1877, a leading botanist wrote that, despite a decade of examining 'different forms of bacteria', he was not persuaded that they could be subdivided into even 'two distinct species'. With this even Joseph Lister agreed. As late as 1881, he declared the evidence for the existence of specific disease-causing germs to be 'entirely untrustworthy'. Lister instead held on to the ancient idea that there are no specific causes of disease, only specific individual responses. Leave a glass of water overnight and then examine a drop the following day under a microscope and it is easy to understand this scepticism. Not only are there just a few general microbial shapes (rod-like, spherical and helical), but as microbes pass through their life cycles, or appear in new environments, they can look and behave very differently. Identifying a bacterium on the basis of what it looks like beneath a cover slip is no simple matter.

The germ theorist's task was made still trickier because he could never be sure that he was using pure cultures of the specific microbes he wished to study. The blood of anthracic sheep, for instance, contains thousands of other bacteria, some harmful, some innocuous. Separating individual bacteria from their myriad contaminants was often impossibly hard. Until that is, Koch provided the solution.

Koch's Finest Hour

In 1881, Koch described a technique he had developed for obtaining pure cultures. Up till then, scientists had used fluid media in which the intermingling of different kinds of bacteria was unavoidable. Koch proposed exchanging this for a solid surface upon which microbial colonies would be far less mobile. His first choice for solid medium was the humble potato, and it worked remarkably well.

The potato was first boiled, then sliced, and

exposed to the air for several hours. A few days of incubation later, close examination revealed the appearance of numerous separate bacterial colonies. 'A few of these droplets', he wrote, 'are white and porcellanous, while others are yellow, brown, grey or reddish and while some appear like a flattened drop of water others are hemispherical or warty.' The crucial point was that these colonies were distinct. As they couldn't get very far on the moist, solid surface of the potato, days later the bacteria were still growing in discrete clusters.

At last, then, here was a method for producing pure cultures. The involvement of the low-tech potato may make this all seem rather prosaic, but the implications of Koch's innovation were immense. From the potato, he soon moved onto gelatine, and finally to agar, an extract of Japanese seaweed suggested by a colleague's wife whose mother had used it to make the perfect jelly. Spreading these media on covered glass Petri dishes, it was now possible to cultivate virtually any bacteria with a greatly reduced risk of contamination.

The notion of using solid media was not solely an idea of Koch's. A scientist working in the same laboratory had also tried it. But it was Koch who made it the touchstone of his biological research. And, armed with this new method, it was his team and the students he trained that forged ahead in the hunt for human disease-causing germs. It is a testament to the importance of this achievement that even Pasteur momentarily overcame his profound Germanophobia and his personal hostility towards Koch to congratulate him. 'C'est un grand progrès, Monsieur', he graciously conceded.

Yet the ability to isolate individual microbes wasn't much use unless the scientist could find them in the tissues and fluids of the diseased. And many germs were so small or transparent that doctors searched for them in vain. Somehow, a method had to be found for making bacteria stand out. The solution to this daunting problem was made possible by the massive explosion in the scale of the European, and especially German, dye industry. By the mid-1870s, hundreds of different textile dyes were available. And, led by Robert Koch, Carl Weigert and one of the future stars of microbiology, Paul Ehrlich, scientists now began to realise that these could be used not only to make bacteria visible, but also to distinguish between different bacterial species.

Not one of the discoveries considered later in this book would have been possible without the development of industrial dyes. Bacteriological research in Germany now marched to the ever accelerating beat of the fastest industrial revolution the world had ever seen; in this, as in so many other respects, the fortunes of science and industry were tightly bound together.

Fehleison's Lucky Break

One of the first successes arising from Koch's new methods fell not to him but to one of his adherents.

Friedrich Fehleison, a surgeon at a Wurzburg clinic who relied on Koch for guidance, was determined to find out the cause of erysipelas, the deadly disease that William Brownrigg had so completely misunderstood in November 1738. But Fehleison knew something that Brownrigg couldn't have known: that germs are always present in biopsies taken from those with the infection. To begin with, however, this knowledge wasn't of much help. Whilst bacteria were indeed found in bloody fluids squeezed from the skin of erysipelas sufferers, such a rich profusion of germs was obtained that it was impossible to say which, if any, caused the disease.

Well primed by Koch, Fehleison began by obtaining sections of erysipelatous skin from corpses and from biopsies. These slices swarmed with different bacteria; but once the skin was spread on a solid gelatine or agar dish, Fehleison was gratified to see that the microbes began to separate and grow in well-demarcated colonies. Following Koch's dogma, he took samples from dozens of erysipelatous patients to see which germs they had in common. He found that only one kind of bacterium was present in all the cases he encountered. Fehleison cultivated this microbe and then injected it into the ear tips of several rabbits. In every case, they developed the spreading lesions typical of erysipelas.

Yet this wasn't quite enough to satisfy all his colleagues. Erysipelas inevitably looks somewhat different in men and rabbits, and Fehleison's critics were quick to point out that he couldn't be sure that he was looking at exactly the same illness. This entirely valid objection momentarily stumped him. Ethical considerations seemed to bar him from turning to human subjects. But how else could he hope to prove his intuition? Things might have stayed at a standstill for years had Fehleison not been exceptionally lucky.

Many of his medical colleagues were convinced that an attack of erysipelas helped in the fight against tumours. This somewhat bizarre belief provided Fehleison with an entirely ethical justification for injecting his cultured microbes into human beings. Having found tumour patients who were agreeable to this treatment, he went ahead. They promptly developed erysipelas and, in doing so, provided Fehleison, and the medical world, with clear evidence for the role of a specific micro-organism in causing a human infective illness. Since the patients luckily survived the direct effects of his intervention, this remained a bloodless revolution.

The Turning Tide

As Friedrich Fehleison found, even as late as 1881 advocates of germ theory still encountered stiff opposition every time they claimed to have identified disease-causing bacteria. However, within little more than a decade and a half this had all changed. During this period, germ theory reached a kind of critical mass. An international effort resulted in so many of the biggest killers being shown to be caused by germs that the public became surprised only if scientists *didn't* find specific microbes in the bodies of the sick. It usually takes something spectacular to make the general population shift its attitudes in the light of new evidence. And this is precisely what the germ theorists delivered between 1881 and 1899. Demonstrating that specific microbial agents cause such devastating diseases as cholera, typhoid, diphtheria, tuberculosis, tetanus, plague, undulant fever and rabies was a massive propaganda victory for the germ theorists. And it is to perhaps the four most startling of these successes that we now turn.

PART VI

THE FOUR BIG ONES, 1881–1899

· Chapter 13 ·

The White Plague

Beautiful but deathly pale, spread languidly upon a chaise longue, her skin stretched taut over prominent cheekbones, coughing gently into a handkerchief speckled with blood. This is how dozens of Victorian operas and novels, from Verdi's *La Traviata* to Hugo's *Les Misérables*, described the slow, excruciating death suffered by victims of tuberculosis. At first glance it seems an oddly romantic way of characterising what is a very horrible disease. But the reason, like so much else in the Victorian period, probably has a lot to do with social class.

Most deadly diseases disproportionately hit the poorer sections of society and, with the snobbery endemic to the period, the élites tended to see in them a measure of moral retribution for idleness and weakness of character. Lacking self-respect, it was implied, the poor hardly minded living among the filth and foul odours that bred illness. The cleanliness and relative good health of their social superiors, in contrast, was seen as the 'outward sign of their inward purity'. But, unfortunately for the élites, tuberculosis was blind to the usual social niceties. With an alarming lack of discrimination, it attacked both rich and poor, good and bad, industrious and idle. Appalled at such apparent injustice, the better-off portrayed tuberculosis as the affliction of the innocent. 'It loves to select the young and beautiful for its victims', wrote one eminent physician. This heavy pathos worked to lessen the stigma by relieving the sufferer from moral condemnation.

But this romantic gloss couldn't really hide the squalid truth. By the mid-nineteenth century, tuberculosis was taking between a tenth and a third of all European and American lives. Rapid weight loss (hence its popular name 'consumption') was followed by severe respiratory problems and the coughing up of coloured sputum. Death then came at varying rates, but was rarely delayed for long. 'That drop of blood is my death warrant. I must die', wrote John Keats as he coughed up his first flecks. And, as autopsies always revealed a mass of white granules in the cavities of the lungs, tuberculosis acquired the chilling epithet 'The White Plague'.

Not least because this affliction was so hard to cure, for most of the nineteenth century it was seen as the classic hereditary disease. Then, in 1865, a French military doctor named Jean-Antoine Villemin claimed to have shown that it could be passed from human sufferers to rabbits and then on to more rabbits; tuberculosis, he announced, is really contagious. But, although Villemin was right, nobody seemed prepared to listen.

This was partly because tuberculosis is only mildly contagious. As such, many doctors failed in their attempts to replicate Villemin's findings. Worse still, during the 1870s Villemin's supporters utterly failed to identify a specific microbe always present in the sputum of consumptives. Despite using exactly the same methods as had revealed the bacteria causing *pébrine*, anthrax and chicken cholera, there was simply no sign of the tuberculosis germ.

The opposition also had a human dimension. As his fellow Frenchman Hermann Pidoux explained, were Villemin correct, 'what a calamity such a result would be! Poor consumptives sequestered like lepers; the tenderness of [their] families at war with fear and selfishness ... If consumption is contagious', Pidoux added, 'we must not say it out loud'. Few people wanted Villemin to be right when proving him so might oblige the government to send those suffering from tuberculosis away from their loved ones, in their thousands, to live out sterile lives in remote sanatoria.

For all sorts of reasons, then, tuberculosis posed a major challenge to the germ theorist.

Koch Conquers All

Once again, however, Koch provided the crucial breakthrough. Two things enabled him to succeed where dozens of others had failed. First, an implacable belief that germs cause disease; and, second, an almost super-human tenacity in the face of repeated disappointments.

Assuming that tuberculosis was indeed caused by a micro-organism, Koch saw that his chief problem was to find a way of making it stand out among hundreds of other bacteria and cells in the sputum of the tuberculous. He turned immediately to the new textile dyes. But despite using all the techniques that had worked so well with other diseases, the tuberculous matter of guinea pigs showed up nothing at all distinctive. By 1880, it was only Koch's refusal to admit defeat that kept him going.

Trying out dozens of different dyes, in different orders and combinations, and at varying temperatures, he at last found a method that revealed the presence of a bacterium that looked like virtually no other. Staining the sputum with methyline blue and then 'washing' it with another dye called vesuvin, he managed to isolate an elongated, wiry but incredibly small germ that seemed to be present in all cases of tuberculosis, in both animals and man. But was it definitely the elusive tuberculosis germ?

Koch now had to satisfy his own postulates. The first step was to grow the germ in an artificial culture. This proved unusually, and exasperatingly, hard. Removing clumps of the mysterious bacillus from tuberculous guinea pigs and spreading them onto his standard media simply didn't work. He then tried new media made of solidified cow and sheep's blood serum. At first, these seemed no better. But, once more, his sheer dogged persistence paid handsome dividends.

With most bacteria, a thriving population normally develops in under 24 hours. But Koch waited patiently for over a fortnight. Only then did a healthy colony of his mystery bacillus finally appear. Persisting with his experiment long after most other researchers would have abandoned it, Koch was at last able to reap his harvest of deadly germs.

Injected into healthy guinea pigs, this elusive bacillus now produced classic consumptive symptoms. Before long, the animals had wasted away, died and were then dissected. Sure enough, their lungs contained the signature granular corpuscles of the white plague. Success now began to come more easily. In the following months, he went on to isolate the same bacilli from dozens of humans and other animals. In each case, he injected them into his guinea pigs which, without fail, developed tuberculosis. Their lungs were later shown to contain masses of the tuberculous bacillus. It was finally time to go public.

Going Public

On 24 March 1882, Koch told a packed meeting of Berlin's Physiological Society what he and his assistants had discovered. Before the session began, he set up a long line of tables with dozens of preparations that reminded at least one attendee of 'a cold buffet'.

Step by step he took his audience through the process by which the tuberculosis bacilli had been found, cultured and tested. 'In the bacillus,' Koch declared, 'we have, therefore, the actual tubercle virus.'

Those who heard him fully understood the enormity of what had been achieved. Koch had solved one of the greatest mysteries in modern medicine and brilliantly succeeded where so many others had failed. 'All who were present were deeply moved and that evening has remained my greatest experience in science', noted Paul Ehrlich many years later. 'It read like a fairy tale to me', wrote an American doctor, having read Koch's paper and undergone a personal conversion experience. In 1865, Villemin had stood almost alone. It was not so now. Much of the medical profession was ready to listen to Koch's claims with a sympathetic mind. Opposition did not completely evaporate and critics were right to argue that some people are predisposed to the illness by poverty or heredity. But in proving that germs are a necessary cause of tuberculosis, Koch had made the germ theory of disease credible and important. Everywhere, from the hospital ward to the parlour room and the public house, the danger of the invisible germ began to capture the public's attention.

In fact, so quickly did word get abroad that some doctors sought to damp down the flames. John Burdon Sanderson, a Professor at University College London, had always been sympathetic to the germ theory. But he now counselled caution and discretion.

'Now the tendency exists to believe germs explain everything', Sanderson complained. For decades, he reflected, the idea that microbes cause illness had had to fight against the fiercest of headwinds. Since Koch's discovery, however, it had 'become one's business to protest with all needful vehemence against the attribution to them of functions which they do not possess'.

In reality, Sanderson had little to worry about. Once the dust settled, it became obvious that germ theorists still had to make their case stick for the big epidemic killers. Here miasma theory continued to hold sway. And not until microbes had also been implicated in such diseases as cholera and typhoid would the germ theory really catch on. Nevertheless, the fledgling science of bacteriology had taken a substantial step forward.

The Tuberculin Debacle

For Koch himself, the tuberculosis story carried an unpleasant sting in its tail. Inspired by Pasteur's success in developing anthrax and chicken cholera vaccines, in 1882 he began searching for a chemical agent that could destroy the tuberculous bacillus inside the body. During the course of these studies, he experimented with a protein substance derived from the tuberculosis-causing bacillus. He called this 'tuberculin'. When injected into healthy guinea pigs, tuberculin had no apparent effect. But when administered to a tuberculous animal or human, a violent local reaction occurred that soon after cleared up. For reasons that are not entirely clear, Koch quickly decided that he had found a cure for tuberculosis. By producing this local reaction, he reasoned, an immune response was triggered at the injection site that would also lead to the eradication of the bacilli from the lungs.

Koch had believed in his hunches before and his tenacity in proving them right had made him a hero of science. This time, though, he was to overreach himself. In 1890, he announced that he had found 'an agent that renders the experimental animal resistant to infection with tubercle bacilli and arrests the disease in infected animals'. Suddenly, an unexpected hope of cure was presented to hundreds of thousands of sufferers. Koch's decision to keep the tuberculin recipe secret only intensified public interest. Berlin's hotels were soon inundated with desperate patients for whom his mysterious 'clear, brown liquid' was the only lifeline. Excitement over Koch's claims was 'at white heat', wrote a medical journalist. Berlin, as one historian has remarked, was turned into a non-denominational Lourdes.

It may be that the German government, eager for one of its scientists to find a cure, persuaded Koch to hype up his tuberculin results. Either way, the consequence for Koch was little short of a personal disaster. In the months and years following his 1890 statement, hundreds of clinical trials showed that tuberculin was of little therapeutic value. The German had believed too readily in what he had wanted to see and had mistakenly perceived some positive early signs as unequivocal successes. The backlash was all the more painful for Koch because he had such a long way to fall. 'The medical world has learnt to believe that any work carried out under the auspices of Professor Koch is thorough and genuine', remarked a correspondent for a British medical journal. Koch just wasn't used to being wrong.

Fortunately, however, his personal life had therapeutic qualities even if tuberculin did not. Having been estranged from his wife for several years, Koch now fell in love with a minor actress whose portrait he had seen whilst sitting for his own. Escaping from Berlin after the tuberculin debacle, he wooed her from Egypt and on his return they married. That Cairo was Koch's choice of retreat when things got rough at home is not without significance, for its narrow, labyrinthine streets had been the stage for another of his greatest successes. Shortly after his triumph with tuberculosis, in Cairo Koch sealed his reputation as the world's foremost microbe hunter by also identifying the cholera bacillus.

· CHAPTER 14 ·

CHOLERA, SUEZ AND PETTENKOFER

In 1855, John Snow had insisted that cholera is caused by a specific micro-organism that passes from victim to victim via infected drinking water. As noted earlier, this bold claim elicited justifiable scepticism. Only by identifying a bacterium that satisfied at least some of Koch's postulates could the medical world be convinced. And, without expertise in microscopical research, Snow had fallen far short of the exacting standards of scientific proof. But where the English doctor had failed, it seemed to many that Koch's and Pasteur's well drilled and equipped teams were bound to succeed. Their big opportunity came in 1883.

In this year cholera struck Egypt. It cut through the poorer districts of Cairo and Alexandria with its customary savagery and fear of its spread caused consternation throughout Western Europe. By August, 5,000 Egyptians were dying every week. And, as Europe braced itself for the disease's onslaught, Britain, France and Germany all rushed expert medical teams to North Africa. The experiences of these expeditions highlight the difficulties involved in gaining acceptance for radical scientific ideas. They also underscore the way in which political circumstances influence how and why medical debates are fought: for the conclusions drawn by the British doctors, who were the first on the scene, were complicated by a surprising factor.

From Cairo to Calcutta

By 1880, almost 80 per cent of the tonnage passing through the Suez Canal was British. In an increasingly competitive world market, the speed with which her ships could travel between East and West was a critical factor in allowing Britain to maintain her economic lead. As the canal slashed the journey time between Britain and India by almost 50 per cent, its military as well as commercial importance was simply immense.

Yet there was a major drawback to having personnel and ships pass so rapidly from one side of the world to the other. With the spices, silks and teas of the Orient could also come disease. Because of this, the crews of merchantmen had traditionally been forced to undergo long periods of quarantine before landing their wares. But quarantine cost Britain time and money, and had long been resented in a nation that depended on speedy maritime communications and uninterrupted trade. In 1882, Britain's economic interest in the Canal led her to make Egypt, in effect, a protectorate of the Empire. Having become the effective suzerains, the British authorities rescinded as many quarantine procedures as they could.

As a result, when cholera hit Egypt in 1883, the British government was placed in a very tight spot. If the disease had been carried on board a ship from India, the colonial power would face the legitimate wrath of thousands of grieving Egyptians. The only way for the British to evade charges of valuing life less than trade was to show that cholera was not a communicable disease. Thus, the government panel that appointed the British medical team was careful to avoid including anyone already sympathetic to the germ theory of cholera. The team selected also lacked experts in microscopy, so that in Egypt it carried out totally inadequate microscopical studies of water supplies and made no attempt to isolate specific microbes.

Partly as a result of this travesty of investigative science, it took the British doctors working under Dr Guyer Hunter only a few weeks to reach their conclusion. According to them, the epidemic was caused by unusual weather patterns 'reactivating' cholera poisons that had lain dormant in the Egyptian soil since the last outbreak in 1865. Cholera had not, therefore, come from India. Nor was it a microbial disease. And there was certainly no cause to expect it to spread any further. There is no reason to think that Hunter and his colleagues had deliberately ruled out the germ theory before starting. The damage had been done by the British government's selection of a team with inadequate microscopical training. And so pleased was the government with Hunter's results that he was promptly made a Knight Commander of the Order of St Michael and St George.

The French and German medical teams arrived in Egypt as the British left, both fully expecting to find the cholera germ. Louis Pasteur helped organise the French expedition, but he remained in Paris and the team was led by Emile Roux. Yet, by the time Roux's men had arrived, the epidemic was on the wane and they struggled to find bodies fit to dissect. Tragically, Louis Thuillier, a bright young prospect in French medical science, contracted cholera from one of the few cases they did find. After a few agonising days, he died. Disconsolate, the French party returned to Paris.

The German expedition, led by Robert Koch himself, laid wreaths at Thuillier's funeral and then settled down to find the microbial culprit. Anatomical studies soon convinced them that cholera is an intestinal disease. And by hunting for distinctive microbes in the intestine, they eventually found their suspect: a comma-shaped bacillus present in the mucus lining of the intestinal tract. Moving their anatomical researches downward to the bowel, Koch's team found this same bacillus in vast quantities in the corpses of cholera victims. These results were extremely promising. But, as the epidemic declined, the Germans too ran out of cadavers.

With little hesitation, Koch's team now took their menagerie of experimental animals on a ship to India, where the disease was much more common. Impressed by what had already been discovered, some British doctors began to feel that an excellent opportunity might have been lost. 'It seems probable that the discovery of the true nature of the virus of cholera will be effected in England's greatest dependency, but not by an Englishman', fumed *The Lancet.* Koch received less begrudging praise from Germany, where Bismarck was relishing the prospect of embarrassing the British.

Shortly after disembarking in Calcutta, Koch's party announced that it had once more identified the comma-shaped bacillus in cholera sufferers. They had also managed to find it in the tanks of water from which many cholera victims had drawn their water. But this was not quite the Eureka! moment that Koch claimed in his official reports. For he repeatedly failed to satisfy his own postulates: over and over again he cultivated the bacillus and injected it into animals that remained stubbornly healthy. This meant that his case rested on only one of his four postulates: a 'constant association' between the bacillus and the presence of cholera.

Clearly, then, this was much less satisfactory than Koch's tuberculosis work. And it gave his rivals scientific as well as political reasons to demur. The British medical élite certainly had some fun attacking 'the eminent Teuton'. But this hostility was driven by scientific logic as much as by national rivalries. After all, even in France and Germany, Koch didn't have it all his own way. To understand why, we need to appreciate the importance of satisfying all four of Koch's postulates.

Extreme Measures

Infecting experimental animals with laboratory cultures often posed a huge problem for the germ theorist. Bacilli like those that cause cholera, typhoid and leprosy are almost uniquely dangerous to man. As a result, many researchers found themselves lodged between the horns of a dilemma. They could test their intuition that animals are resistant to what they believed to be a disease microbe only by resorting to morally indefensible human experimentation. But if they didn't do so, their critics could quite justifiably accuse them of propping up the germ theory with a face-saving fudge. This difficulty is perhaps most apparent in the case of Norwegian leprosy research.

For centuries, doctors assumed leprosy to be a hereditary disease. Then the Norwegian doctor Gerhard Armauer Hansen found distinctive rod-shaped bacteria in the skin nodules of his leprous patients. Isolating them using some of Koch's methods, he next injected the bacteria into dozens of rabbits. To his great frustration the animals showed no signs at all of the disease. So, assuming that rabbits are immune, he turned to members of his own species.

At his trial in May 1880, Hansen explained how he had made an incision in the eyelid of a female suffering from leprosy with a sharp knife earlier used to slice through the leprous nodules of another patient. His objective was to see if his female patient developed the particular strain of leprosy from which the first ailed. As his second patient was already leprous, Hansen argued that the common good justified human vivisection. In the event, however, infection was not transmitted. And, because he had subjected a non-consenting patient to a painful, dangerous and ultimately futile procedure, Hansen was stripped of his position in Bergen's leper hospital.

It was later shown that Hansen's bacillus was indeed the cause of leprosy. But his failure to infect his patient's eye with leprosy highlighted yet another problem for the germ theorist. Even where disease usually can be spread between members of the same species, the presence of bacteria is not always correlated with the illness itself. Thus, doctors trying to replicate Koch's and Pasteur's results often found that pure bacterial cultures of tuberculosis or anthrax had no effect at all on some members of species ordinarily susceptible to the disease. Critics of Koch's germ theory of cholera likewise pointed out that only a few of those who drank from tanks containing infected water actually succumbed.

Part of the problem was that Koch and Pasteur often spoke as if ingesting germs was not only a necessary, but also a sufficient condition for developing infectious disease. This was an assertion contradicted by many of their own experiments. Yet, rather than just modifying germ theory to take into account varying susceptibilities to disease, some critics went a good deal further. The leading German hygienist Max von Pettenkofer, for example, didn't deny that germs are involved in the cause of infectious illness. But he claimed that they aren't nearly enough. An epidemic, Pettenkofer argued, requires three things: a specific infectious particle; moist, porous soil containing decaying organic matter; and a toxic substance produced by the combined effects of the particle and the soil.

In an attempt to prove the validity of his theory, in 1884 Max von Pettenkofer wrote to Koch and asked him to send some cholera vibrios (bacteria that look like curved rods). Having complied with this request, shortly after Koch was stunned to receive the following letter:

> *Herr Doctor Pettenkofer presents his compliments to Herr Doctor Professor Koch and thanks him for the flask containing the so-called cholera vibrios, which he was kind enough to send. Herr Doctor Pettenkofer has now drunk the entire contents and is happy to be able to inform Herr Doctor Professor Koch that he remains in his usual good health.*

Herr Doctor Pettenkofer survived only because his high stomach acidity neutralized the cholera vibrios. But the almost insane bravado of Pettenkofer's demonstration excited huge interest in his own hybrid of germ and miasma theory.

A British doctor called Edward Klein was actually the first person to undertake this auto-experiment. In July 1884, Klein led a British medical team to India to look into Koch's work. Before his expedition departed, another doctor, Timothy Lewis, travelled to Marseille where cholera had struck and he reported being unable to find Koch's comma-shaped bacillus in many of the cases he examined. A few months later, Klein's India team fully endorsed Lewis' findings. Koch's constant association postulate, he argued, was not always fulfilled. D. D. Cunningham, another British doctor based in India, wholeheartedly agreed. Into the 1890s, Cunningham clung to the view that local soil conditions determine whether or not cholera breaks out and to where it spreads. Cunningham also insisted that Koch's comma bacillus was only one of several microbes capable of causing the disease. Once more, of course, British commercial interests were at stake. But there is no strong evidence that this swayed either Klein's or Cunningham's judgements.

Hubris and Haffkine

If this battle was not yet won by the time Koch left Calcutta, great advances had been made. Furthermore, even Koch's most ardent critics, the loyal followers of Pettenkofer, were convinced that one couldn't contract contagious disease without germs being involved. Thus by the mid-1880s the germ theorists had achieved at least a partial victory. But

Koch would not be satisfied until it was universally accepted that if an individual ingested the comma bacillus only their natural defences could save them from death. This meant disproving Pettenkofer's notion that local soil conditions are as important in producing cholera as the germ itself. As Koch's supporters soon realised, this was no easy task. Ironically, though, it was made a lot easier by the hubris of their opponents.

Publicly swallowing glasses of cholera bacilli and suffering few ill effects became almost a stock-in-trade for Pettenkofer and his Russian-born ally Elie Metchnikoff. Yet in 1893, a M. Jupille volunteered for one of these trials and all but died from a classic choleric attack. A few months later, Metchnikoff learned that cholera bacilli had been found in a river near Versailles and that none of the townspeople had fallen victim to the disease. So Metchnikoff persuaded several local people to ingest solutions of this microbe. All of them fell very ill and, shortly after, one tragically died. Incidents such as these gradually convinced scientists, not least Metchnikoff himself, that Koch's bacilli alone were enough to induce the disease.

A consensus that the comma bacillus is the *only* microbe that can cause cholera also emerged during the 1890s. But again, at first things didn't go Koch's way. A Spanish scientist, Jaime Ferran y Clua, reported that he had inoculated thousands of his countrymen against cholera using a pure culture of the comma bacillus. His claims of an impressive success rate brought a French Commission to his doorstep. Unfortunately, he then withheld vital information and his vaccine was declared useless. Ferran's vaccine probably did work, but its repudiation by a high-ranking foreign team brought it and the germ theory of cholera into temporary disrepute.

At this juncture, the baton was taken up by a Russian-born scientist called Waldemar Haffkine. Haffkine was born in the prosperous Black Sea port of Odessa into a large Jewish family. Vehemently left-wing, at University he was embroiled in revolutionary politics. Then, during the savage anti-Jewish pogroms that followed the 1881 assassination of Tsar Alexander II, he also became involved in attempts to organise Jewish resistance. Attracting the unfriendly attention of the secret police, he began spending more and more time in jail. Haffkine was living dangerously, and he seemed destined for the gallows.

The Russian, however, was ambitious. He wanted to cut a swathe in the scientific world and knew that as a well-known insurgent he had no future in his native land. So, he fled to Geneva, where he took up bacteriological research. A few years later, in 1890, he moved to Paris where he started work under Roux and began trying to develop a cholera vaccine. This was an exciting project for a young scientist with a lot to prove. Quite apart from the prospect of saving countless lives, Haffkine knew that developing a comma bacillus vaccine would win the admiration of some of the world's leading scientists. Should he have needed further encouragement, the Prince of Siam promised to 'erect a statue' to him if he succeeded. 'Let's see now what we are capable of', he jotted in his notebook the evening of the Prince's visit.

Haffkine's main problem was developing and testing potential vaccines when laboratory animals seemed immune to the disease. But, early on, he made something of a breakthrough. If he mixed cholera bacilli in test tubes with rabbit serum, only the most virulent and hardy bacteria managed to survive their encounter with the serum (later it would be shown that this is because of the presence in the serum of antibodies). If Haffkine then repeated the process using ever-larger quantities of rabbit serum, he eventually obtained a very potent culture of comma bacilli indeed. Injected into several guinea pigs, Haffkine experienced his first real sense of triumph when some of them fell ill and died. Yet, although this showed that laboratory animals could be infected with the bacilli, only a minority of rabbits and guinea pigs were actually affected. Encouraged and frustrated in roughly equal measure, he persevered through 1891 and early 1892 without making further progress.

A word of advice from his boss ended this impasse. Emile Roux informed Haffkine of a discovery made by Koch's colleague, Richard Pfeiffer, who had found that cholera bacilli injected through the lining of the guinea pig's abdominal wall almost invariably proved fatal. Haffkine seized his chance. He injected comma bacilli into the abdominal cavity of a guinea pig.

When it died, he syringed some of the 'exudates' from its abdominal cavity and transferred them to the abdomen of another guinea pig. As Pasteur had previously shown, passing bacteria through dozens of animals increases the strain's virulence. After 30 passages, Haffkine's comma bacillus was almost unfailingly lethal. He dubbed it his 'exalted' strain.

Next, Haffkine set about trying to attenuate his pure culture. This proved fairly simple. Exposing the bacilli to air, to heat or to chemical agents weakened them sufficiently for use as vaccines. Animals first inoculated with these bacilli nearly always survived a subsequent injection of the otherwise fatal exalted strain. Soon after, Haffkine found that attenuating the bacilli wasn't even necessary. If he injected exalted bacilli subcutaneously, the animals acquired immunity. Somehow, this route gave the body the chance to develop an effective immune response before the full-blown disease could manifest itself.

Having obtained these results, Haffkine took the momentous step of injecting himself with his attenuated bacilli. Pain at the injection site and a raised temperature were his only symptoms. A week later, Roux gave his junior colleague a full-blown cholera injection. Haffkine survived after a spell of only moderate discomfort. So did three of his Russian friends who were vaccinated the same week. Then, two months later, while holding a pipette of cholera exudates in the air, Haffkine felt several drops of the 'viscous, slightly salty' liquid fall into his mouth. This was a very potent mixture and he hesitated for a moment before gulping down the microbe-rich fluid. His immune system having coped with this massive dose, Haffkine's confidence in his vaccine never wavered again.

Now he sought a chance to prove his vaccine's worth on a grander scale. Russia turned him down flat. And so, perhaps worrying that it might have to fulfil its earlier promise, did the kingdom of Siam. Thus Haffkine turned to India and, thanks to the mediation of the former viceroy Lord Dufferin, by the spring of 1893 he was on board a vessel bound for the sub-continent. Once in India, though, he met a barrage of opposition. Much of it came from local people, who resented the painful injections and the days of labour lost to the vaccine's unpleasant side effects. But doctors too joined the chorus of complaints. They quite legitimately accused Haffkine of over-confidence, since the disease he had induced in laboratory animals looked nothing like cholera. How, they asked, could he be sure that he was dealing with the same illness?

A temporary decline in the incidence of cholera in 1893 meant that Haffkine was given no opportunity to answer his critics. Then, in early 1894, it broke out in a crowded and insanitary suburb, or bustee, of Calcutta. Rushing to the scene, Haffkine quickly set up shop and had soon vaccinated 116 of the 200 bustee residents. All 116 escaped the disease, whereas several of the unvaccinated died. Moving into the

Illustration 6: Using Haffkine's vaccine to protect the Third Gurkhas against cholera in India during 1893. Reproduction, 1894, of a wood engraving. Source: The Wellcome Library, London.

tea country of Assam, he continued his vaccination programme. There he managed to vaccinate 20,000 people. One of his colleagues later calculated that of those vaccinated, only 2 per cent died of cholera. Of those who had not submitted to vaccination, the mortality rate fluctuated between 22 and 45 per cent. This was a ringing endorsement for the vaccine and for Koch's germ theory of cholera. And, having learned of Haffkine's labours, the German remarked with delight, 'the demonstration is complete'. By that stage, the majority of doctors, even in Britain, were inclined to agree.

One of the most terrifying diseases of the nineteenth century was not only understood, but it could be prevented. Whatever lingering embarrassment Koch might have felt after the tuberculin affair now evaporated with the belated recognition that in 1883 he had indeed discovered the cause of cholera.

· CHAPTER 15 ·

PASTEUR'S GATEKEEPER

Deaths from rabies and drowning in quicksand are to be found in Victorian fiction with a frequency that belies their rarity in the real world. But it is largely because of the sheer intensity of suffering that precedes death in each case that they came to symbolise the grimmest kinds of demise. Not least because it is the more drawn out, most would agree that rabies is the worse of the two.

Before this disease was brought under control, a bite from a rabid animal condemned the victim to many weeks of agonising waiting before he or she could tell whether or not the disease had been transmitted. If unlucky, the first sign of the onset of rabies was a fever and a general sense of malaise. Muscle aches, vomiting and a sore throat followed shortly after. With the fever now at around 105°F, the victim became subject to severe spasms, seizures, hallucinations, confusion and extreme sensitivity to bright lights, sounds and touch. Finally, as the disease laid waste to vital areas of the nervous system, breathing and swallowing became very difficult. The characteristic foaming at the mouth began as the body tried to expel bloody sputum that could no longer be swallowed. And, although desperately thirsty, sufferers shrunk with hydrophobic terror from any liquids offered to them. Soon after, utterly disoriented, violent and scared, he or she slipped into a coma and, within hours, stopped breathing. Perhaps the worst aspect was that all this was known from the outset. For the rabies victim, death was virtually inevitable.

This sorry litany helps us to understand why Pasteur elected to tackle rabies. It may have been an extremely uncommon infection, but he realised that conquering perhaps the most terrifying and deadly disease known to man would represent a coup for the germ theorist of immense proportions. Since it had a death rate of 100 per cent, Pasteur also saw that he could resort to human vivisection with ethical impunity. If the victims were going to die anyway, what was to stop him testing his experimental vaccines on them first? But Pasteur also had some personal demons to exorcise.

In Autumn 1831, an 8-year-old Pasteur had witnessed a terrible chain of events unfold that he would remember for the rest of his life. A wolf in an advanced stage of rabies was seen at the edge of the wooded foothills of the Jura Mountains. With teeth covered in bloody saliva laced with rabies virus, it mauled anything in its path. Leaving the woods, the wolf found its way into Villers-Farlay and savagely attacked several villagers. Increasingly crazed, frightened and aggressive, it next wandered into Pasteur's home town of Arbois, where it bit a dozen or so more. These victims rushed to the centre of town where the local blacksmith cauterised their wounds with a red-hot iron. The young Louis heard the screams as the iron came into contact with wounded flesh. Over the following weeks, the town was plunged into mourning as eight of the wolf's victims suffered ghastly deaths. Few medical researchers could have had stronger personal reasons for wanting to get to grips with this terrible disease.

A Flying Start

Pasteur's drive to find a rabies bacillus was always doomed to fail. We now know that rabies is a viral disease, so there was no bacillus to be found. To make matters yet more difficult, its causative agent can be seen only with the aid of a modern electron microscope. Fortunately, this did not stop Pasteur's team making progress. Nor would it afterwards be held against them. With rabies, the reputation of the germ theory quickly came to hinge on whether or not his team could produce an effective vaccine.

So, in early 1884 Pasteur, working alongside Roux, began trying to make dogs immune to the disease. They soon found that passing matter taken from infected spinal cords through the brains of one rabbit after another had the effect of increasing the virulence of the rabies virus. After a few dozen passages, the potency plateaued out and infection always appeared in about six days. This gave Pasteur and Roux a rabies strain of predictable intensity, and they termed it the *virus fixe*.

Quite by chance, while conducting these experiments they also noticed a paradoxical phenomenon. If the *virus fixe* obtained from serial passages in one species was then injected into a different species, the virus behaved as if it had been much weakened. For instance, rabies virus passed through dozens of rabbits or monkeys could give dogs complete immunity to subsequent injections of even the most virulent forms of the disease. As these vaccines worked in only 60 per cent of cases, they were not totally reliable. Nonetheless, Pasteur and Roux knew that they were tantalisingly close to having one that would work every time.

Experimenting alone, Roux next decided to try using high temperatures to attenuate the viruses. Removing a rabbit spinal cord riddled with rabies, he suspended it in a glass jar and sent it to be incubated in the communal oven. Just as the new batch of trays, including Roux's jar, was being placed inside, Pasteur happened to walk by and immediately spotted the jar. Once informed that it was Roux's, he didn't need to be told any more. Returning to his own workspace, Pasteur copied Roux's method with the innovation of a few rocks of caustic potash that dried the air and quickened the process of desiccation.

Roux was furious when, a few days later, he found that his idea had been poached. The temper, for which he was famous, flared up and he stormed out of the laboratory, slamming the door behind him. A rapprochement was not long in coming, however, and both men collaborated in testing the dried out spinal cords on another set of dogs. The animals were first injected with dry, and therefore less virulent, spinal cord matter. Then, over the course of several days, fresher and more potent spinal cords were used. The hope was that immunity would be gradually built up so that the dogs would survive a final, typically fatal dose of *virus fixe*.

The First Human Trials

Soon after work began, Pasteur sent word around Paris' hospitals asking that he be notified if any patient were admitted with rabies symptoms. Almost immediately he heard of the case of a Parisian called M. Girard, who had been bitten by a wandering dog. Although the man's wound had healed, on 1 May he rushed to Paris' Necker hospital in a state of considerable agitation. Unable to bring himself to drink water or wine, suffering from a severe headache and with his legs shaking uncontrollably, M. Girard feared the worst. The attending physician, Dr Rigal, fired off a telegram to Pasteur's laboratory and both Pasteur and Roux were quickly on the scene. A hurried conference followed, during which it was agreed that

M. Girard was indeed suffering from rabies and that he would certainly die.

It didn't need a moral philosopher to justify testing the spinal cord vaccine on a man who seemed sure to die an agonising death. The dog studies had proven encouraging. And even if lots more were needed before the prototype vaccine could be judged safe, it looked to be M. Girard's only hope. So, a few hours later, Pasteur and Roux returned from the laboratory with a syringe-full of attenuated rabies matter, and a subcutaneous injection was immediately administered. That night, the two men went back to the hospital with another full syringe. In the interim, however, the hospital authorities had got cold feet. Concerned as to the ethical status of Pasteur's tests, they terminated M. Girard's treatment and he was abandoned to his fate.

Over the next few days, thoughts of an excruciating death tormented M. Girard. Yet only a few weeks later, on 22 May, he was discharged from hospital, cured. No one knows what eventually happened to M. Girard, and the reasons for his recovery remain a mystery. Perhaps the most likely explanation is that he was not suffering from rabies at all, but from a temporary psychosomatic disorder induced by the fear that he was.

Nevertheless, the incident must have impressed on Pasteur and Roux the importance of completing their animal trials as soon as possible. Certainly, just a few days after M. Girard's discharge, they began a new series. Ten dogs were injected daily, from 28 May till 9 June, with rabid spinal cords of increasing freshness. Between 3 June and 18 June, another ten were subjected to the same procedure.

A few weeks into these trials, Pasteur had a second chance to test his vaccine on a human patient. He was called to see an 11-year-old girl called Julie-Antoinette Poughon. The child had been bitten on the upper lip by her puppy and there was no doubt at all that she was suffering from rabies. Neither Pasteur nor the medical staff had any reservations about the propriety of commencing an immediate course of attenuated virus injections. Sadly, though, it was too late. After only two shots, the girl died.

Although this wasn't the most auspicious of results, both Pasteur and Roux recognised that inducing immunity once the disease had already set in was almost certainly impossible. So, just two days after Julie-Antoinette's death, they stepped up their laboratory work. Another ten dogs were injected with rabies matter: Pasteur and Roux were now going all out to test their latest vaccine.

A Boy from Alsace

On Monday 6 July, just five weeks after the first ten dogs had been inoculated, a 9-year-old peasant boy from Alsace was brought by his anxious mother to Pasteur's doorstep. Having heard of the great scientist's research, this distraught mother implored Pasteur to save her son's life. Young Joseph Meister had been bitten severely and deeply by a rabid dog. Pulled from beneath it, he was found to be covered with the creature's blood and saliva. The dog's owner cornered and killed it, and then cut open its stomach. It was found to contain a mass of hay, straw and woodchips, typical of a rabid animal.

Moved by the mother's harrowing account and by her obvious distress, Louis Pasteur was confronted with a cruel dilemma. Girard and Poughon had both been positively diagnosed with rabies before he had proposed administering rabies vaccine. In both cases, he had therefore been fully justified in making use of a largely untried therapy. But by no means all people bitten by rabid dogs develop rabies. In fact, the majority, even many of those bitten deeply like Joseph Meister, never contract the disease at all. This wouldn't have been an issue had Meister appeared at Pasteur's door just a few weeks later. Yet, in early July 1885, his dog experiments were far from completed. Sure, of the first ten dogs, all ten were healthy despite having received fresh rabies matter in their final injection over a month earlier. This gave grounds for optimism, but the incubation period for rabies can be several weeks, so Pasteur couldn't be certain that these first ten dogs were really in the clear. About the others, he could only speculate.

As such, Pasteur knew that if young Meister were free of rabies and the trial vaccine failed, then he might end up giving the boy a full dose of the disease and condemn him to a death that he would not otherwise have suffered. On the other hand, he also realised that there was a high chance that Meister was indeed carrying the rabies virus and that the disease's long incubation period meant that, if he acted fast, he could confer immunity on him before the virus had a chance to act. Few scientists can have faced such a difficult ethical choice, yet it seems that Pasteur knew what he was going to do from the outset.

To his chagrin, Roux shrunk from the chance of proving the vaccine's efficacy. Unconvinced that the dog experiments had been carried far enough, Roux strongly recommended that Meister's fate be left to chance. Pasteur was adamant that this was the wrong approach. Thus, despite 'deep concern', on 6 July he began giving the boy more and more virulent injections of infected spinal column. In all, thirteen inoculations were carried out over ten days. The first of these contained rabies matter so desiccated that it had little or no effect when injected into the brains of rabbits. But the last few inoculations contained sufficiently virulent viruses to kill anyone who lacked immunity.

Sleep could not have come easily to Pasteur through late July and early August. In October, however, when he addressed the Academy of Science, he was able to tell his audience that Joseph Meister's health did not 'leave anything to be desired'. The boy was fit, happy and profoundly grateful to his apparent saviour. Pasteur's boldness, some might say foolhardiness, had brought him a fabulous success. By this time, the dog experiments he had begun in June had also proven thoroughly successful; not one of the 40 inoculated dogs succumbed to the final injections of highly virulent rabies virus.

Shortly after Meister was confirmed well and rabies-free, Pasteur received a letter from the mayor of a village close to his home town. Mayor Perrot told him of the gutsy actions of a 15-year-old shepherd boy called Jean-Baptiste Jupille. A few days before, a powerful dog with 'suspicious gait' had approached a group of young children with every intention of attacking them. Jean-Baptiste knew a rabid dog when he saw one and, seizing his whip, he charged it. The dog lunged at him and managed to sink its teeth into his left hand. But, after a fierce struggle, Jean-Baptiste wrestled the dog to the ground, tied his whip around its muzzle and beat it to death with one of his metal-tipped clogs. As Pasteur later remarked, this was a feat of uncommon 'bravery and cool-headedness'. During this heroic action, however, the dog had inflicted several deep bites.

The boy's story was all the more tragic because he was his family's sole breadwinner. Shortly before this incident, Jean-Baptiste's father had lost his arm, and as a result his job, in an industrial accident. Now, it seemed, his son was destined to die of rabies.

This time Pasteur had no reservations in agreeing to help. But the boy's parents were uneasy. Before they put Jean-Baptiste on the next train bound for

Illustration 7: A photograph of the statue outside the Pasteur Institute depicting Jupille's struggle with a rabid dog. Source: The Wellcome Library, London.

Paris, they had to be assured by the Mayor that 'their son was lost unless they accepted the generous offer of M. Pasteur'. On 20 October 1885, almost as soon as he arrived, the course of increasingly virulent inoculations commenced. Pasteur's sense of urgency was entirely justified, since this young man didn't reach Paris until six days after receiving his bites. Nevertheless, a month later the danger had passed. Jean-Baptiste was in excellent health. And the efficacy of the vaccine, and the validity of the germ theory of rabies, were clear to virtually all.

Luck Comes to the Bold

At the close of Pasteur's 1885 speech before the Academy of Science, in which he described his rabies work, the audience responded with rapturous applause. A member of the French rabies commission got to his feet and declared that saving the lives of Meister and Jupille had put 'the finishing touch on the glory of our illustrious colleague and adds the most distinguished lustre to our country'. It was a sentiment with which much of the world agreed. Pasteur now assumed a celebrity on a par with the very greatest heroes of the history of science: his place in its Pantheon was guaranteed. The chancy and debatable ethical decisions he had made were soon forgotten, Roux returned to the fold and the scientific world widely accepted Joseph Lister's colourful judgement that the remaining critics were simply 'ignoramuses'.

Other doctors subsequently experienced real difficulties in using Pasteur's methods to produce the rabies vaccine. Nevertheless, its efficacy was proven over and over again in the following years. By the time of Pasteur's death in 1895, an extraordinary 20,000 people had been successfully immunised against rabies at centres all around the globe. As the accolades poured in, so did financial contributions. In response to Pasteur's request for the funds needed to build a vaccine research centre, thousands of people, from dukes to tradesmen to schoolchildren, donated money to the cause. The now world-famous Pasteur Institute opened its doors in November 1888. Far more than a monument to its founder, who was later interred therein, it rapidly became the world's foremost centre for the teaching and study of microbiology.

Fittingly, Joseph Meister became the caretaker of the Pasteur Institute, a job he held for the remainder of his life. According to several accounts, he committed suicide in 1942, rather than obey the instructions of a German officer to open Pasteur's tomb.

· CHAPTER 16 ·

TYPHOID FEVER

In the middle of November 1861, Albert, the Prince Consort, developed a serious fever and took to his bed. The Royal physicians were summoned, but their patient quickly adopted an oddly fatalistic attitude. 'I do not cling to life', he said to a distraught Queen Victoria, adding: 'I have no tenacity for life.' With Albert seemingly resigned to his own death, the doctors feared that his downcast frame of mind might seriously worsen his condition and cause the Queen insufferable distress. So, in a series of upbeat bulletins to the press, they deliberately understated the severity of Albert's illness.

On 7 December, however, Albert developed a rose coloured rash on his skin. There was now little doubt that he had typhoid fever and that his life was in the balance. Still, in order to safeguard the royal couple's peace of mind, positive statements continued to emanate from the Palace. At last, after a month of very severe diarrhoea, dehydration and fever, Prince Albert died. A stunned nation learned for the first time that he had been seriously ill. Traumatised, Victoria imposed the regime of mourning she would observe for the rest of her life.

The medical literature in the weeks following Albert's death was fixated on typhoid fever. Many doctors still denied that it was a specific disease with a specific cause. Few believed bacteria to be in any way involved. Even its mode of transmission was the subject of intense disagreement. But it was in 1861 that the Bristol doctor, William Budd, published *Typhoid Fever*, a ground-breaking work in which he argued that typhoid is a specific disease caused by a microbe spread via contaminated water.

Budd's book initially generated more heat than light. Yet, with his encouragement, during the 1860s doctors gradually came to appreciate that dirty water is the source of the infection. And, as in so many other cases, antiquated sanitary arrangements were quickly shown to have played a key role in Albert's death. Typhoid is 'the very offspring of foul sewers and ill-drained dwellings', concluded the *British Medical Journal*. At the same time, powerful evidence suggesting that typhoid is a specific disease was also emerging. Karl Wunderlich, a German physician, showed that typhoid can always be distinguished from other kinds of fever by the temperature of the sufferer. It probably was, therefore, a distinct kind of infection.

Nevertheless, neither Budd nor any of his immediate contemporaries had found a typhoid germ. This is hardly surprising. For in addition to ingenuity and stubborn determination, it would take highly sophisticated laboratory techniques to prove that typhoid, one of the horrors of the Victorian age, is a microbial disease. Yet, by 1900 this merciless scourge was all but conquered, and proof that typhoid fever is caused by bacteria then played an essential role in consolidating the germ revolution.

Hunting the Bacillus

Carl J. Eberth was a morbid anatomist based in Zürich. Between 1880 and 1881, he worked on sliced sections of spleen and intestinal lymph nodes he had taken from patients dead of typhoid fever. With the help of appropriate staining techniques, he identified in them masses of rod-shaped bacteria. This was the first promising evidence that typhoid was associated with a specific microbe. Encouraged, Eberth proceeded to dice up the internal organs of forty more typhoid victims in the hope of finding his rod-shaped bacteria again. In eighteen of these cases, his search paid off. This was mildly disappointing, but all bacteriologists knew that not finding a germ didn't necessarily mean that it wasn't there. So Eberth took the next logical step.

He sliced, stained and examined the spleens and intestinal lymph nodes of 24 people who had most definitely not died of typhoid fever. This time Eberth was pleased to find absolutely no trace of his distinctive bacilli. On a much smaller scale, the same results were also reported by two British scientists, Joseph Coats, who worked in Glasgow, and G. F. Crooke of Leeds. All three men fell far short of providing proof that typhoid is a bacterial infection. But, by the close of 1881, germ theories of typhoid had at last been brought within the pale of scientific probability.

The lead was now taken by Georg Gaffky, yet another of Koch's students. Using more sophisticated staining and microscopical techniques, he identified Eberth's rod-shaped bacilli in 26 out of 28 typhoid cases. This impressive result fully justified his subsequent attempts to culture the organism. To do this, Gaffky took the entire spleen of a typhoid victim, sterilised its outer surface by washing it in mercury chloride solution, and sliced through it with a clean knife to expose the spleen's inner cavity. Next, he used a sterilised platinum wire to scrape off tiny fragments of tissue, which were then streaked across a nutrient gelatine plate. Finally, this plate was covered and left at room temperature.

Forty-eight hours later, Gaffky had a pure culture of what is now known as the Eberth-Gaffky bacillus. He was also able to demonstrate that this was a bacterium hitherto unknown to medical science. Instead of forming round, coloured or opaque colonies, it grew as a thin, almost invisible slime upon the surface of his media. For years, this peculiarity would be the primary means by which scientists distinguished it from all other bacilli.

But, at this high point, Gaffky's luck ran dry. By now it was known that typhoid is an intestinal infection. So in order to show that his bacillus might be the cause of typhoid, he needed to find it in the stools of its victims. This proved beyond the technical competency of Gaffky's laboratory. The stool samples he smeared on gelatine plates contained large quantities of what are known as saprophytic bacteria, which have the effect of liquefying gelatine and making it impossible for other bacteria to form colonies. Worse still, all Gaffky's attempts to transmit typhoid to experimental animals were abject failures. To most germ theorists it seemed likely that typhoid was exclusively a human disease, just like cholera and leprosy. But critics insisted either that Gaffky had isolated the wrong bacteria or, less charitably, that he was on the wrong track entirely.

At the Berlin International Medical Congress in 1890, Koch commended Gaffky's work. But he added that any claims to the effect that the Eberth-Gaffky bacillus was the cause of typhoid could 'only be received with legitimate doubt'. Despite intensive efforts in numerous laboratories and in many different countries, even as late as 1896 the medical world was divided on what to think. Humans are, as Gaffky supposed, among the few species susceptible to typhoid fever. But proving that this was why Koch's postulates couldn't be satisfied was no simple matter. In short, Gaffky's frontal assault had failed and a new, less direct, line of attack had to be devised.

After years of stalemate, in April 1894, the same year as Haffkine's triumph in India, a breakthrough finally occurred. In the Vienna laboratory of Max von Gruber, two British students, Herbert Durham and Albert Grunbaum, had taken blood serum from typhoid sufferers and poured it into test tubes containing solutions of the Eberth-Gaffky bacillus. They were stunned to observe the bacteria immediately forming into viscous clumps. Further studies revealed that the serum did not have this effect on any other kind of microbe. Durham and Grunbaum quickly surmised that, inside their host, Eberth-Gaffky bacilli trigger the production of a specific kind of immune response, the function of which is to disable the invading bacteria by clumping them together. Whatever caused the clumping had been present in the serum that had been injected into the test tubes of Eberth-Gaffky bacilli and had had the same effect there as inside the body.

Albeit rather oblique, this observation provided first-class evidence that the Eberth-Gaffky bacillus really did cause typhoid fever. Otherwise how could one explain why the serum of typhoid sufferers caused this single kind of bacillus to form into clumps? Unfortunately for Durham and Grunbaum, the French scientist Georges Fernand Widal announced his development of a similar method before they published their results. They forfeited their priority, but, with the widespread adoption of the Widal test, the deadlock in typhoid research was decisively broken.

Wright and the Ladysmith Siege

With the uptake of the Widal test throughout Europe and America, resistance to the specific germ theory of typhoid receded. As testing centres were established in many large cities, the association between typhoid symptoms and the presence of the Eberth-Gaffky bacillus became obvious to most. Just four drops of blood taken from the ear lobe, mixed with a solution of typhoid bacteria, quickly showed whether or not someone was infected. Nevertheless, there was still scope for arguing that the supposed typhoid bacilli were merely secondary invaders that made themselves at home in hosts already suffering from typhoid. Only the development of a typhoid vaccine by the British scientist Almroth Wright finally demolished this objection.

Argumentative, acerbically witty and the inspiration for the arrogant Sir Colenso Rigeon in Bernard Shaw's *The Doctor's Dilemma*, Wright was nonetheless a highly skilled scientist. Based in the Pathology Department of the Army Medical College in rural Netley, he closely followed Haffkine's development of an anti-cholera vaccine. Inspired, in 1885 he set about finding ways to weaken the Eberth-Gaffky bacillus. Thanks to another critical discovery by Richard Pfeiffer, this didn't take him long. Pfeiffer showed that if typhoid bacteria are killed – rather than merely weakened – by heat treatment and then injected into the body, an immune response is still produced.

These graves at Intombi, where the hospital camp outside Ladysmith is situated, speak only too eloquently of the ravages caused by enteric fever among our troops in South Africa. In Ladysmith, no doubt, the epidemic was aggravated by the scarcity of good food. The officer who sent the photograph states that at the time of writing there were 800 patients at Intombi. From the returns issued by the War Office it appears that while 2,355 officers and men have been killed during the war and 575 have died of wounds, no fewer than 2,803 have died of disease

GRAVES OF ENTERIC PATIENTS IN LADYSMITH

Illustration 8: Rows of graves of those – mostly the unvaccinated – killed by typhoid and other enteric fevers at Intombi, Ladysmith. Halftone *c*.1900, after a photograph (anonymous). Source: The Wellcome Library, London.

The potential for attenuating bacteria by killing them had been discussed before. But most doctors had clung to the assumption that only a living microbe has what it takes to trigger the body's mysterious immune processes. Pfeiffer shot this notion out of the water and Wright was only too glad to confirm his findings. Several batches of typhoid bacilli were thoroughly heated and then put through the Widal test. Sure enough, the characteristic clumping reaction still took place. Wright now knew that he had the makings of a viable vaccine.

As finding people on whom to experiment presented a major challenge, Wright and his staff very bravely stepped into the breach. Through 1897 and early 1898, dozens of different dilutions of dead typhoid bacilli were injected into the staff of the Pathology Department. Serum was then drawn from their arms a few days later and the Widal test performed. In most cases, the clumping reaction occurred: humans could indeed mount an immune reaction against dead bacilli. Yet, mere evidence of the presence of antibodies was hardly proof that immunity had been acquired. As Durham and Grunbaum had shown, even the blood of those killed by typhoid usually passed the Widal test. Thus, had this test been available to assay Prince Albert's blood in December 1861, for instance, there is little doubt that the bacteria would have formed thick, unwieldy clumps in his serum too.

As a result, if Wright was to be sure that his vaccine

really did protect against typhoid, he had to find someone willing to receive a second injection containing live microbes. This involved a high level of danger, for Wright really had no evidence that the first injection granted full immunity and he may well have smoothed over the risks involved in order to induce one staff member to undergo this critical test. Luckily for Wright, this mute inglorious hero of the germ revolution experienced no ill-effects. So far, so good. But it was still with premature confidence that Wright offered his services as soon as typhoid broke out in the Maidstone area of Kent in 1897.

Caused by contaminated mains water, the Maidstone typhoid epidemic killed nearly 150 people in a few months. A fund for 'stricken Maidstone' was set up and it attracted donations from all over Europe. Queen Victoria, celebrating her fiftieth year on the throne, sent a large donation and a message of sympathy, no doubt tinged with personal sadness. The Maidstone epidemic was particularly shocking because outbreaks of typhoid were much less common than they had been. The alarm generated seems to have overwhelmed normal scruples about human vivisection, and Wright was given approval to proceed on the strength of what would now be considered woefully inadequate experimental evidence.

Soon after the start of the epidemic, typhoid struck at the Barming Heath mental asylum. Out of a staff of about 200, twelve had already contracted the disease and 84 others were sufficiently worried to volunteer for vaccination. Wright happily obliged. A few months later, as the epidemic receded, he and his medical team took stock. It transpired that none of their human guinea pigs had died, whereas four of the unvaccinated staff members had succumbed. This sample size was far too small for proper conclusions to be drawn. But some of Wright's superiors gladly gave him their benediction.

Among those applauding were high-ranking officials in the government of India. They offered Wright the chance he now desperately needed to try out his vaccine in a country in which typhoid was virtually endemic. With its dogmatic aversion to public spending, the British government refused to fund his expedition. Undaunted, Wright set out for India in 1898 as a member of the India Plague Commission, and proceeded to test his vaccine on a large scale. Over the course of several months, he gave his vaccine to any officer or soldier prepared to receive it. The results convinced most army doctors of the vaccine's effectiveness. Dozens wrote to the viceroy expressing their satisfaction at their troops' immunity from a fever that, in the past, almost invariably hit them on long marches and when encamping in lowland areas. Impressed, the British government was persuaded to vaccinate any member of the army willing to undergo the procedure. Unfortunately, however, the statistics Wright obtained from India were not of a high calibre. Largely for this reason, many of his medical colleagues remained unimpressed.

In 1899, the start of the Boer War gave him another chance. Relentless petitioning by Wright and his allies ensured that the troops sent to the Cape were strongly encouraged to receive his anti-typhoid inoculations on the long journey south. Winston Churchill, then a reporter sailing to South Africa aboard the *Dunstlar Castle*, described a typical scene:

> *Inoculation ... proceeds daily. The doctors lecture in the saloon. One injection of serum protects; a second secures the subject against attack. Wonderful statistics are quoted in support of the experiment. Nearly everyone is convinced. The operations take place forthwith, and the next day sees haggard forms crawling about the deck in extreme discomfort and high fever. The day after, however, all have recovered and rise gloriously immune.*

Churchill himself was sceptical about the value of immunisation. Why he objected isn't entirely clear. Perhaps he was among the many libertarians who considered Wright's activities to be on the slippery slope towards compulsory vaccination, something they found deeply objectionable. But regardless of this, Wright was now on the verge of his greatest triumph.

In 1899, roughly 21,000 Boers encircled a garrison of about 12,000 British troops holding the town of Ladysmith. With no supplies coming in, conditions rapidly worsened and by the third month of the siege, most of the town was subsisting on horsemeat and tiny allocations of rice. To make matters much worse, because the town's water supply had become contaminated by human and equine excrement, typhoid struck hard. Ultimately, far more soldiers died of this disease than from the Boers' copious shellfire.

But, as statistical data subsequently indicated, the Ladysmith troops had not been equally susceptible. Roughly 17 per cent were lucky enough to have been vaccinated. Among this group, the mortality rate from typhoid fever was just one in 213. The unvaccinated suffered far more grievously. One in 32 of them were killed by the Eberth-Gaffky bacillus. These statistics were highly impressive. The damage to Britain's reputation, and the military losses suffered as a result of the Boer War were severe, but at least events in Ladysmith served to convince most doctors of the validity of the germ theory of typhoid.

Nevertheless, after Ladysmith anti-vaccinationists remained hostile to the army's wish to inoculate all military personnel. Even after the outbreak of World War I, stalwart defenders of the freedom of the individual derailed attempts to vaccinate compulsorily all new recruits. The army therefore resorted to subtle means of getting its way. In September 1914, Lord Kitchener announced that only men who had first received the jab would be allowed to fight abroad. Such was the desire of the troops to get at the enemy that before long as many as 80 per cent of those dug into the trenches of Belgium, Holland and France were fully protected against typhoid. By 1916, virtually the entire army could boast immunity.

Germany and France had also developed anti-typhoid vaccinations, but neither had been pushed with the determination mustered by Wright and his supporters. Both countries began vaccinating their soldiers later, and were much less thorough. The result was an unanswerable argument in favour of vaccination. Throughout the war, there were only 7,000 British cases of typhoid compared with 125,000 among the French and 112,400 among the Germans. For all the other agonies they bore, the British troops of World War I were at least protected against one of the biggest killers of soldiers in human history. It is interesting to reflect that in a war of attrition, in which marginal advantage was everything, Wright's vaccine might well have been the difference between victory and defeat.

· CONCLUSION ·

A NEW SCIENCE

'A new science has been born', gloated Louis Pasteur in June 1888; 'it has caused a veritable revolution in our knowledge of virulent and contagious diseases'. The Frenchman was not exaggerating. The germs responsible for anthrax, cholera, tuberculosis, leprosy, diphtheria and gangrene had already been found. The next ten years would see the discovery of the microbial causes of plague, scarlet fever, tetanus, typhoid fever, pneumonia, gonorrhoea and cerebrospinal meningitis as well. Virtually every year between 1879 and 1899, scientists unlocked the secret of another important infectious disease; never before had medicine witnessed progress so rapid or so fundamental.

In fact, many doctors felt overwhelmed by the punishing pace of change. Those who had been taught from textbooks in which germs were never mentioned, now had to start almost from scratch. So much had to be learned and unlearned that many clung on to some parts of older medical theory, interpreting the new in light of the old. But for the new generations of doctors emerging from professional medical schools, the study of infectious disease began with Pasteur, Koch, Lister and Roux. Virtually everything that went before suddenly seemed obsolete: Hippocrates was the author of an oath and nothing else of value. Medicine had ceased to be an art and had become a fully fledged science.

In reality, of course, the germ revolution hadn't emerged out of nowhere. Modern germ theory is the culmination of more than 2,000 years of observation and investigation. Nor did it immediately translate into massive health benefits. In the Western world, civilian deaths from big epidemic killers were already in steady decline thanks to the vast improvements in sanitation, drainage and water-quality made possible by Victorian public health reformers and civil engineers. Standards of nutrition were also on the rise, and with them people's resistance to infectious disease. As Pasteur himself noted in 1888, the germ revolution over which he had presided produced a transformation in human 'knowledge' more than in treatment or prevention.

But despite these caveats, it is hard to overestimate the importance of the fact that, for the first time in human history, doctors properly understood what causes infectious illness. At last they had a solid platform on which to build, and tangible benefits almost immediately accrued. The germ theory revealed that mankind is besieged by disease agents invisible to the naked eye. Far from being restricted to obviously insanitary people and places, they might be found anywhere. Scrupulously aseptic surgery, wide-ranging hygienic reforms in catering and food production, the chlorination of mains water and pasteurised dairy products were among the revolution's earliest and most important progeny.

A recognition of the deadly threat posed by unseen germs also had an enormous public impact. Around the turn of the century, studies found potentially lethal bacteria on clean fabrics, food and household objects, even children's toys boxed up and then taken out for the amusement of the next generation. Such findings shocked the public into adopting new attitudes about the prevention of sickness. Some of these bordered on the hysterical. Kissing, touching and sharing clothes and linen were strongly and openly discouraged. In many parts of the United States, even the common cups used during communion were threatened: cashing in on the growing fear of germs, Sanitary Communion Companies offered worried vicars patented sets containing individual glass cups, disinfectant solutions and serving trays. Some germ evangelists went further still and forcefully recommended the abolition of the handshake.

More sober and less fetishist reactions to the invisible germ included large-scale public campaigns to persuade those with infectious diseases to avoid coughing and sneezing in others' faces and to dispose of contaminated material in responsible ways. Hundreds of rural mansions were converted into sanatoria, where those with contagious diseases were nursed in isolation. Hand-washing using powerful disinfectants became routine practice. Listerine, the first mass-produced mouthwash, appeared on the chemist's shelf. And for those who could afford such things, the ornate Victorian water closet gave way to the Spartan, almost puritanical, modern bathroom. The white china toilet surrounded with smooth, tiled walls and floors became the new ideal.

But it was the second-generation offspring of the germ revolution that most emphatically fulfilled Pasteur's 1888 dream of also 'destroying germs'. As he predicted, knowledge born of the germ theory gave previously unimaginable power to prevent and to fight infectious disease. From the earliest synthetic drugs, the sulphonamides developed by Paul Ehrlich and Sahachiro Hata, to the antibiotic revolution that has saved countless lives since the 1940s, the cure of bacterial infection has become a routine medical affair. The vaccinations we receive as children or before travelling to tropical countries are also the lineal descendants of the one Pasteur and Roux developed after finding the microbial agent of chicken cholera in 1880. How many of us would have gone to an early grave, were it not for the formulation of vaccines like the anti-tuberculosis jab?

Indeed, only the term 'revolutionary' can convey a proper sense of the magnitude of the change that medical practice has undergone. Until the mid-twentieth century, cynicism about the efficacy of

Illustration 9: One of dozens of similar posters produced by the British Ministry of Health during and immediately after World War II, warning of the danger of spreading disease through the failure to trap germs in handkerchiefs. Lithograph after Henry Bateman. Source: The Wellcome Library, London.

medical remedies was the mark of the educated, self-aware doctor. After all, who but a quack could claim that the pharmacopoeia's myriad tonics and coloured pills did anything other than delude the patient into parting with their cash? 'I firmly believe that if the whole materia medica [i.e. medical drugs], as now used, could be sunk to the bottom of the sea, it would be all the better for mankind, – and all the worse for the fishes', wrote the American doctor, Oliver Wendell Holmes, in 1880.

By the end of World War II, doctors had much less need for bluffing. The drugs they dispensed in ever-increasing quantities actually worked. The amorphous category of untreatable diseases that had to be borne, with whatever stoicism the victim could muster, had shrunk dramatically. Antibiotics had put microbes on the back foot in a spectacular fashion. And, thanks to improved medical care, nutrition and sanitation, the biggest microbial killers in the Western world by 1974 – influenza and pneumonia – ranked way behind heart disease, cancer, strokes and accidents in the mortality stakes. Bronchitis, the only other microbial illness making the top ten, barely scraped in. For the first time in human history, most Western patients with infectious illnesses could approach doctors in the confident expectation of a cure. Today, those who cannot be helped feel very badly let down. What a difference 100 years can make!

Bibliography and Further Reading

There are several books that cover in more detail the major themes of this book. Among the best are Roy Porter's comprehensive *The Greatest Benefit to Mankind: A Medical History of Humanity from Antiquity to the Present* (London: HarperCollins, 1997), William F. Bynum's *Science and the Practice of Medicine in the Nineteenth Century* (Cambridge: Cambridge University Press, 1994), Robert Reid's *Microbes and Men* (London: BBC Books, 1974), W. D. Foster's *A History of Medical Bacteriology and Immunology* (London: Heinemann Medical, 1970), Hubert A. Lechevalier and Morris Solotorovsky's *Three Centuries of Microbiology* (New York: McGraw-Hill, 1965), Nancy Tomes' *The Gospel of Germs: Men, Women, and the Microbe in American Life* (London: Harvard University Press, 1998) and the highly idiosyncratic *The Microbe Hunters* (New York: Harcourt, Brace and Co., 1926, reprinted 1996) by Paul de Krief.

Part I

For further information on William Brownrigg, see Jean E. Ward and Joan Yell (eds) *The Medical Casebook of William Brownrigg, MD, FRS (1712–1800) of the Town of Whitehaven in Cumberland* (London: Wellcome Institute for the History of Medicine, 1993). Eighteenth-century medical thought, and its social and intellectual underpinnings, is described in Christopher Lawrence's *Medicine in the Making of Modern Britain, 1700–1920* (London: Routledge, 1994), Nicholas Jewson's provocative 'Medical knowledge and the patronage system in 18th century England', *Sociology*, 1974, vol. 8, pp. 369–85 and Guenter B. Risse's 'Medicine in the age of enlightenment', in *Medicine in Society: Historical Essays* (Cambridge: Cambridge University Press, 1992), edited by Andrew Wear, pp. 149–95.

Part II

The themes of contagion, infection, miasma, early public health efforts and attempts to combat childbed fever are described in: Margaret Pelling's entry 'Contagion/germ theory/specificity', in the *Companion Encyclopedia of the History of Medicine* (London: Routledge, 1993), Irvine Loudon's *The Tragedy of Childbed Fever* (Oxford: Oxford University Press, 2000), Roy Porter's *The Greatest Benefit of Mankind* and Lester S. King's *The Medical World of the Eighteenth Century* (Huntington, NY: R.E. Krieger Publishing Co., 1971). For discussion of the discovery of microbes and debates about spontaneous generation, see John Farley's *The Spontaneous Generation Controversy from*

Descartes to Oparin (Baltimore: Johns Hopkins University Press, 1977). Many of the original papers are translated and reproduced in Thomas D. Brock's *Milestones in Microbiology: 1546 to 1940* (Washington, DC: ASM Press, 1999). Bynum's *Science and the Practice of Medicine* provides an excellent introduction to Paris medicine. For the myths and realities of the John Snow story, see *The Lancet* article entitled 'Map-making and myth-making in Broad Street: the London cholera epidemic, 1854', by Howard Brody, Michael Russell Rip, Peter Vinten-Johansen, Nigel Paneth and Stephen Rachman, all of whom are academics at Michigan State University (*The Lancet,* 2000, vol. 356, pp. 64–8).

Part III

By far the best and most reliable account of the life and work of Louis Pasteur is the late Gerald Geison's *The Private Science of Louis Pasteur* (Princeton: Princeton University Press, 1995). René Dubos' *Louis Pasteur, Free Lance of Science* (New York: Scribner, 1976) remains a good biography. K. Codell Carter's article, 'The development of Pasteur's concept of disease causation and the emergence of specific causes in nineteenth-century medicine', in the *Bulletin of the History of Medicine,* 1991, 65, pp. 528–48, contains many valuable insights. For Joseph Lister, see: Christopher Lawrence and Richard Dixey's 'Practising on principle: Joseph Lister and the germ theories of disease', in Lawrence's *Medical Theory, Surgical Practice: Studies in the History of Surgery* (London: Routledge, 1992); Lindsay Granshaw's '"Upon this principle I have based a practice": the development and reception of antisepsis in Britain, 1867–90', in the book *Medical Innovations in Historical Perspective* edited by John V. Pickstone (Basingstoke: Macmillan, 1992). Richard Fisher's *Joseph Lister, 1827–1912* (New York: Stein and Day, 1977) remains a good biography.

Part IV

Pasteur's research on silkworms is analysed in Antonio Cadeddu's 'The heuristic function of "error" in the scientific methodology of Louis Pasteur. The case of the silkworm diseases', *History and Philosophy of the Life Sciences*, 2000, vol. 22, pp. 3–28. For the life and works of Casimir Davaine, see 'Casimir Davaine (1812–1882): a precursor to Pasteur', by Jean Théodoridès in *Medical History*, 1966, vol. 10, pp. 155–65. For Robert Koch, see Thomas D. Brock's biography, *Robert Koch: a Life in Medicine and Bacteriology* (Washington, DC: ASM Press, 1999). Also useful are Geison's *The Private Science of Louis Pasteur* and Cadeddu's 'Pasteur et le choléra des poules', *History and Philosophy of the Life Sciences*, 1985, vol. 7, pp. 87–104.

Parts V and VI

Porter's *The Greatest Benefit to Mankind*, Bynum's

Science and the Practice of Medicine in the Nineteenth Century, Reid's *Microbes and Men*, Brock's *Robert Koch* and Lechevalier and Solotorovsky's *Three Centuries of Microbiology* all provide detailed descriptions of the discovery of the microbial nature of tuberculosis, cholera, rabies and typhoid. Michael Worboys explores the British context and reception of these discoveries in *Spreading Germs: Disease Theories and Medical Practice in Britain, 1865–1900* (Cambridge University Press, 2000). For the history of tuberculosis, see also David Barnes, *The Making of a Social Disease: Tuberculosis in Nineteenth-Century France* (University of California Press, 1995). Mariko Ogawa's 'Uneasy bedfellows: science and politics in the refutation of Koch's bacterial theory of cholera', *Bulletin of the History of Medicine*, 2000, vol. 74, pp. 671–707, and Ilana Löwy's 'From guinea pigs to man: the development of Haffkine's anticholera vaccine', *Journal of the History of Medicine and Allied Sciences*, 1992, vol. 47, pp. 270–309 shed further light on the cholera research of Koch and Haffkine. Geison's *Private Science of Louis Pasteur* provides a fine revisionist account of Pasteur's rabies experiments. Lastly, Anne Hardy's '"Straight back to barbarism": antityphoid inoculation and the Great War, 1914', published in the *Bulletin of the History of Medicine*, 2000, vol. 74, pp. 265–90, traces the opposition to typhoid inoculation during the early twentieth century.

Harvey's Heart
The discovery of blood circulation
ISBN 1 84046 248 5 £5.99

Turing and the Universal Machine
The making of the modern computer
ISBN 1 84046 250 7 £5.99

Dawkins vs Gould
How the battle of evolutionary theory is still fought over, some 130 years after Darwin first announced his theory
ISBN 1 84046 249 3 £5.99

Available in hardback

Watt's Perfect Engine
Steam and the age of invention revealing the warts-and-all James Watt
ISBN 1 84046 361 9 £9.99

Constant Touch: A Global History of the Mobile Phone
A fascinating exploration of the mobile phone
ISBN 1 84046 419 4 £9.99

Lovelock and Gaia
How Gaia theory came about and its long struggle to gain acceptance
ISBN 1 84046 458 5 £9.99

Kuhn vs Popper
The struggle for the soul of science
ISBN 1 84046 468 2 £9.99

Sex, Botany and Empire
The story of Carl Linnaeus and Joseph Banks
ISBN 1 84046 488 7 £9.99

The Autobiography of Charles Darwin
ISBN 1 84046 503 4 £9.99

Introducing series science titles (paperback £9.99)

Introducing Artificial Intelligence	1 84046 463 1
Introducing Chaos	1 84046 078 4
Introducing Consciousness	1 84046 115 2
Introducing Darwin and Evolution	1 84046 153 5
Introducing Einstein	1 84046 060 1
Introducing Environmental Politics	1 84046 159 4
Introducing Evolution	1 84046 265 5
Introducing Evolutionary Psychology	1 84046 043 1
Introducing Fractal Geometry	1 84046 123 3
Introducing Genetics	1 84046 120 9
Introducing Mathematics	1 84046 011 3
Introducing Mind and Brain	1 84046 084 9
Introducing Newton/Classical Physics	1 84046 158 6
Introducing Quantum Theory	1 84046 057 1
Introducing Relativity	1 84046 372 4
Introducing Science	1 84046 358 9
Introducing Stephen Hawking	1 84046 096 2
Introducing Time	1 84046 263 9
Introducing The Universe	1 84046 068 7